T0091973

LIVING WITH ROBOTS

LIVING WITH ROBOTS

WHAT EVERY ANXIOUS HUMAN NEEDS TO KNOW

RUTH AYLETT AND PATRICIA A. VARGAS

FOREWORD BY NOEL SHARKEY

THE MIT PRESS CAMBRIDGE, MASSACHUSETTS LONDON, ENGLAND

The MIT Press would like to thank the anonymous peer reviewers who provided comments on drafts of this book. The generous work of academic experts is essential for establishing the authority and quality of our publications. We acknowledge with gratitude the contributions of these otherwise uncredited readers.

This book was set in Stone Serif and Avenir by Westchester Publishing Services. Printed and bound in the United States of America.

Library of Congress Cataloging-in-Publication Data

Names: Aylett, Ruth, 1951- author. | Vargas, Patricia A., 1969- author. | Sharkey, N. E. (Noel E.), writer of foreword.
Title: Living with robots : what every anxious human needs to know / Ruth Aylett and Patricia A. Vargas ; foreword by Noel Sharkey.
Description: Cambridge, Massachusetts : The MIT Press, [2021] | Includes bibliographical references and index.
Identifiers: LCCN 2020035369 | ISBN 9780262045810 (hardcover)
Subjects: LCSH: Robots--Popular works. | Artificial intelligence--Popular works.
Classification: LCC TJ211.15 .A95 2021 | DDC 629.8/92--dc23
LC record available at https://lccn.loc.gov/2020035369

10 9 8 7 6 5 4 3 2 1

To Peter Aylett, whose example led me into computing, and Rob Jones, whose love and comradeship made it possible for me to have four children and be an active robotics researcher.—RSA

To my partner, for all the support and unconditional love. You always make me feel safe and give me comfort.—PAV

CONTENTS

FOREWORD ix
Noel Sharkey
INTRODUCTION xiii

1 WHY ARE WE SO SCARED OF ROBOTS? 1
2 APPEARANCE: WILL THEY LOOK LIKE US? 21
3 MOVEMENT: WILL THEY LIVE WITH US? 41
4 SENSES: WILL THEY BE AWARE OF US? 61
5 THE LOST ROBOT: COULD THEY KNOW WHERE THEY ARE AND HOW TO GET HOME? 77
6 TOUCH AND HANDLING: COULD I SHAKE HANDS WITH A ROBOT? 91
7 COULD ROBOTS BE AIS? 109
8 COULD ROBOTS LEARN TO DO THINGS FOR THEMSELVES? 127

9 COLLABORATING ROBOTS: COULD THEY WORK AS PARTNERS OR GROUPS? 145

10 EMOTIONS: COULD ROBOTS HAVE FEELINGS? 161

11 SOCIAL INTERACTION: PETS, BUTLERS, OR COMPANIONS? 179

12 SPEECH AND LANGUAGE: WOULD WE BE ABLE TO TALK TO THEM? 201

13 SOCIETY AND ETHICS: COULD A ROBOT HAVE MORALS? 217

ACKNOWLEDGMENTS 237

NOTES 239

INDEX 275

FOREWORD

You might have picked up this book and are thinking, "Oh, no, not another book on robotics and artificial intelligence." But before you put it down, let me tell you that it is a breath of fresh air and a book that you ought to read if you, like me, don't want to be fed the excrement of bulls.

We've all experienced hype—extravagant and unfounded claims usually propounded for commercial purposes. I still remember my bitter disappointment over the ray gun that I really wanted for Christmas in 1957. It turned out to be a torch with three colored lights. On the positive side, it taught me a lot about the hype of advertising and made me think much more critically about my toy choices.

But this overselling seems so innocent compared to the hype today surrounding artificial intelligence (AI) and robotics. There have always been false claims and overexaggeration about the nature and abilities of robots. As early as 1927, a robot created by the US appliance company Westinghouse was widely reported in the world's media as a new domestic slave that would be able to carry out all household chores within the next ten years. This was the first example of robot hype, but certainly not the last.

Robots tug strongly on our natural tendency to project human or animal attributes onto inanimate objects. Such anthropomorphism (or zoomorphism) has long been exploited by sales personnel and designers. We see it particularly in car design, with everything from aggressive "gnashing-teeth" models to cuddly family cars. Look around and see how many "faces" you can find. You may not realize that you are projecting animal characteristics onto objects, but those qualities affect how you see them, the way you interact with them, and which products you buy.

With robots, the anthropomorphic tug is powerful. They have the appearance of independent movement—like an animal, but a lot stiffer. Scientists have suggested that robots form a new perceptual category between inanimate and animate. And this is largely through deception. A number of humanoid-style robots on the market have been manufactured to look cartoonishly cute, with cuddly turns of the head or some other appealing gesture. They often get demonstrated at commercial fares, conferences, festivals, and even on TV with a hidden operator controlling them. They may also have scripted speech that can provide answers to questions known in advance. Sometimes the operator just talks through the robot in a robot voice.

Where's the harm? It's all a bit of a laugh and good fun, isn't it? The harm lies in the creation of a mythology that grossly overestimates the capabilities of the machines. It pushes us into the possibility of a new category of being.

And the massive overexaggeration around AI amplifies the overblown claims of robotics. After all, what else would be controlling our robots?

Since its beginning in the 1950s, AI has had strong ambitions to create a computer program that could think like a human. With intellectual inflation, by the 1980s, "thinking" wasn't enough. Now computers had to be superintelligent. They would quickly become much smarter than us and take control, either to benefit us like a kind and just god or to kill us all off like Terminators.

I don't know if this could happen, and neither does anyone else. But I have not seen a shred of scientific evidence to make me consider the possibility. Even the incredible DeepMind program, AlphaGo, that has beaten the world's best player in the highly complex game of Go, is strictly limited to the game. It cannot do anything else; it does not understand what a game is or even that it is playing one. It doesn't care if it wins or loses. It is no more a step toward reaching superintelligence than walking upstairs is a step toward reaching the moon.

The hype, the mythology, and the general silly talk weren't much of a problem in the twentieth century. But that has all changed. Now the overhype and the overestimation of the capabilities of robots and AI are leading us into dangerous territory that has negative impacts on our society.

We can cite some obvious examples, such as the use of computer programs to make decisions that affect our lives: decisions such as who will get a mortgage, a passport, a loan, or a job, and even whether you will get bail if arrested. It turns out that the rapid development and widespread use of these "decision algorithms" were a terrible mistake. In the rush to roll them out, they were not tested effectively, and they have turned out to be racially, ethnically, and gender biased. The oversubscription to the algorithms of injustice is just one of the consequences of people's misunderstanding of the limitations of the technology.

Another, more worrying example is that the same sort of algorithm is being developed for use by the major military powers. They are hell-bent on developing robot weapons such as fighter jets, tanks, warships, and submarines. Despite AI's inability to understand or comply with the laws of war, these robot weapons are set to go out on their own, find their own targets, and kill them without human supervision. How crazy is that? And it all grows directly out of the hype.

Yet robots and artificial intelligence have great potential to benefit humanity as we negotiate through the new terrain of climate change and global pandemics. This book makes a strong contribution to our understanding of the capabilities and limitations of the technology. If you prefer fairy-tale stories about robots, this is not the book for you. It cuts through the hype to tell you what is important about robotics in our lives now and what they cannot do.

Noel Sharkey, June 2020

INTRODUCTION

Headlines tell us robots with human or even superhuman capabilities are just around the corner. Economic reports identify robots as a serious threat to jobs, claiming that fifteen million jobs in the United Kingdom are at risk. So much of what we read and see in the media reinforces these fears. Should we worry? Maybe research into intelligent robots should be banned or highly regulated?[1] Can we live with robots at all? If so, how?

In this book, we suggest that the way to understand what robots can genuinely contribute—good and bad—is to get to grips with what they really are: the technologies that give them bodies, power sources, mobility, sensing capacities, the ability to set and meet goals. We will see that our everyday environments, so simple for us, often seem almost impossibly difficult for robots. We will demystify the wilder assertions and explain what is easy and what is hard when it comes to giving robots the capabilities they need to function like us and around us.

Let's start by thinking about the language we use to talk about robots. After all, language is what we use to understand what robots are capable of and whether we should be anxious about their development.

One of the pioneers of artificial intelligence, Marvin Minsky, talked about what he called *suitcase words*.[2] By this he meant words that carry not one but many different meanings, often related to the context in which they are used. His examples were *consciousness*, *emotions*, *memory*, *thinking*, and *intelligence*, many of which appear in discussions about robots. Consciousness is often what people mean when they talk about robots becoming self-aware or coming to life. *Emotion* is another word we use to distinguish between humans and things that look human but are not. *Thinking* and *intelligence*, or often *superintelligence*, are words that underline people's anxieties about the future development of robots.

Minsky suggests that trying to arrive at a single definition for these words is much too hard, but we do need to *unpack* them, or examine what we mean exactly in context. The "too hard" reflects the fact that even specialist researchers in the relevant field do not agree on the definition. When it comes to *intelligence*, many specialists argue that the term is a misnomer, and intelligence does not exist at all as one thing, never mind one measurable thing. If human intelligence is itself a suitcase word, then what should we make of artificial intelligence as an expression? What exactly do we mean by it? Certainly the fact that we use these important words as suitcases holding many meanings does not make a realistic assessment of robots and robotics any easier.

This book is all about unpacking, about looking at robots concretely as human artifacts rather than placeholders for our anxieties. In chapter 1, we start by looking at around 2,500 years of robot-like devices and explore the roots of our present panics. In chapter 2, we consider what the word *robot* really means to roboticists. Then, in successive chapters, we look at how we give robots some basic capabilities: movement in chapter 3; senses in chapter 4. In chapter 5 we consider how robots know where they are and how they can navigate to where they are supposed to be; in chapter 6 we look at grasping and touching, as well as the use of robot-inspired prostheses for amputees and other people with physical disabilities. In chapter 7 we unpack that big word *intelligence* and explain how specialists define it. Their ideas may come as a surprise to readers outside the field.

In chapter 8, we look at another suitcase word, *learning*, and explain what machine learning does and does not offer for a robot. Chapter 9 takes us beyond the single robot to swarms, cooperating robots—from robot soccer to search and rescue—and humans and robots working together. In chapter 10 we explain why models of emotion may help robots to interact more smoothly, and in chapter 11 we trace how far the development of social robots has progressed. In chapter 12 we discuss the capability everyone wants in a robot, speech and language interaction. Finally, in chapter 13, we return to the big issues of ethics and social impact: killer robots, sex-bots, and whether robots really will take your job.

Along the way, we will see how robots measure up to humans and other creatures. In trying to produce competent robots, we roboticists come to appreciate just how amazing living things are.

1

WHY ARE WE SO SCARED OF ROBOTS?

We begin this story with a kind of sex robot.

The classical Roman author Ovid, in his *Metamorphoses,* stories of transformation, tells of an ancient Cypriot named Pygmalion. Pygmalion was what we would describe as a misogynist: he thought all women full of terrible vices and would have nothing to do with them. But he was also a sculptor and made a lifelike statue of an idealized woman. Then he fell in love with his own creation. He moved it into his bedroom and spent hours kissing and caressing it and looking into its eyes. Eventually he got so desperate that he went to a temple of Venus and prayed that the statue would become his wife. This being fiction, of course, Venus obliged, and Pygmalion's statue came to life, becoming Galatea.[1]

If not a robot, Galatea is certainly close to what we would now call an android, a word invented in the early eighteenth century.[2] The idea that the ideal woman is the passive creation of a man can still be seen in current-day sex robots—mechanized versions of the more traditional life-size blow-up doll. But this idea is more widely present in the disturbing tendency for humanlike robots in the present day to be represented as young women.[3]

If we can trace some versions of actual robots back to the story of Pygmalion, then a second thread in our ideas about robots can be traced to a different set of myths. These also focus on animated statues but relate strongly to metalworking technology and emphasize strength and ability to labor.

An example of these stories is Talos, a giant man of bronze, created by the mythical Greek smith-god Hephaestus. Talos was made to protect Europa, one of the young women impregnated by Zeus. Zeus, in the form of a large white bull, carried off Europa and left her, pregnant, on Crete. Talos circled the island three times a day to prevent anyone else kidnapping—or perhaps rescuing—her. Hephaestus was said to have developed similar metallic figures to help in his workshop. Daedalus, the mythical inventor at the court of King Minos, also in Crete, is another subject of myths about lifelike metal figures. Daedalus's figures could also speak because he imbued them with quicksilver, or mercury. In some stories, his living statues had to be tied down, or else they would wander off on their own.[4]

Here the emphasis is on a fearsome strength and invulnerability rather than beauty and sexual attractiveness. But the crucial difference between these stories and that of Pygmalion is that technology, rather than some mysterious life-giving force, is involved. The mythical prototypes might have been built by gods or legendary inventors, but these stories clearly implied that metalworking skills, rather than divine inspiration, were the key issue.

These two themes, of magical life-giving force and mind-boggling technology resulting in creations that have human or even superhuman functionality, come right down to the present day. These entities were not called robots, but they fill the same hole in the human psyche. After all, as the science fiction writer Arthur C. Clarke pointed out, "Any sufficiently advanced technology is indistinguishable from magic." The boundary between what could be built by sufficiently skilled artisans or inventors, and mysterious magical powers, was often blurred in these myths. Even today, accounts of robots also blur the boundary between actual technology and mysterious and awe-inspiring accomplishments, and as we will see, robots in film have a lot in common with mythology.[5]

The ancient world went beyond merely telling stories of robot-like artifacts to actually constructing them. Alexandria was a center of this technology. The earliest of its documented engineers, a man named Ktesibios, lived in the third century BCE. He was probably the first head of the Library of Alexandria and wrote extensively about his work. His own writing did not survive, but it was so influential that it was cited by other authors, whose works we do still have.

In myths, the power source animating statues or metal warriors can be skated over or vaguely allocated to gods. An engineer has to solve the problem in the real world. Of course, there was no electricity, the most usual engineering choice in the modern day. The devices Ktesibios built used pumps to compress air or cause water flows: pneumatics and hydraulics. His manual explained how to construct the pumps and described a compressed-air catapult and a water-powered organ. But he also included pneumatic birds that sounded the hours on a water clock. His water clock design remained the most accurate available until the invention of pendulum clocks in the seventeenth century.[6]

Animated birds were popular choices for engineers building robot-like devices, since they made an impressive display at a monarch's court. Artificial trees with singing birds at the court of the Caliph of Babylon figured in songs from the time of Charlemagne, and at the court of the Byzantine emperor during the same period. They were installed at the tenth-century Palace of the Tree in Samarra, also in the Arabic world.[7]

You could say that robot-like devices had three purposes in the ancient world: sex and labor were two of them, and spectacle, or entertainment, forms the third, with mechanical singing birds offering a good example. This idea too has come down to the present day, with animatronic dinosaurs, robotic toys, and robots as a marketing device with which to impress customers in banks or museums.

The ancient world also had much more spectacular robot-like devices than singing birds. One of Alexandria's kings, Ptolemy II, commissioned a grand procession in 279/78 BCE in which a gigantic female statue called Nysa—twelve feet high when seated—rose to its full height, poured a libation of milk, and sat down again. We have no extant design for Nysa, but recent analysis suggests a complex mechanism of cams and weights, and a sprocket chain or gear wheels.[8]

The last of the Alexandrian engineers, Hero, wrote designs for two large-scale projects containing multiple moving figures. One was a mobile shrine to Dionysus with a miniature god and his maenads, and the other was a miniature theater that staged a small play.[9] The theater rolled onto the stage by itself, and tiny figures enacted a multiscene tragedy, after which the artifact rolled off again. It was driven by a large descending weight whose speed was regulated by a sand clock. The pull of the weight provided the force for the links and gears driving the figures.[10]

When we read contemporary or later accounts from the ancient world of these artifacts, the tone is one of wonder and fascination. What we do not see is any concern that these robot-like devices pose a threat or can somehow substitute for humans. Of course, one-off devices made by a few extraordinarily talented engineers could never be seen as a "robot army." Neither do we see any speculation that such engineering projects might take people's jobs away. In those societies, slaves carried out most of what we would see as real work. Nobody discussing the animated devices of that period sees them as able to perform as slaves. The idea just does not come up. Indeed, the steam engine invented by Hero of Alexandria did not lead to automation or factory production but was applied to spectacles like opening temple doors. While we are just as attuned to spectacle and as easy to impress with our own robot technology, our very different social and economic context means we view robots in our own day very differently from those early devices.

Certainly the use of hydraulics and pneumatics limited the capabilities of the animated figures of the ancient world. They had to be closely coupled with their unwieldy and fixed sources of power. This meant that the mobility of Talos or of Daedalus's fictional assistants was beyond the technology of the day. This began to change once mechanically driven clockwork was reintroduced into western European societies in the 1300s. One of the most famous collections of animated figures of that period, at the chateau of Hesdin, in northern France, was still hydraulically driven.[11] However, figures animated by clockwork became increasingly popular.

Spring-driven clockwork was certainly known in the ancient world, but much less discussion of it has come down to us than of water and compressed air as sources of power. The one definite example of a spring-driven mechanism is not an animated figure but the Antikythera device, found in a shipwreck off the Greek island of that name in 1901. This device can be described as an analog computer, since it uses models of the solar system to predict astronomical events. The team that analyzed the device suggested that it could model the movement of the sun and moon through the zodiac and thus predict both lunar and solar eclipses, and theorize it was wound with a hand crank.[12] It is unlikely that the device was a one-off, and probable that spring-driven clockwork was used for other mechanisms, though we have no strong evidence for clockwork-driven figures in the ancient world.

The metalworking sophistication and engineering knowledge needed to build clockwork mechanisms were lost at some point in western Europe and slowly returned from the 1300s onward via the Islamic world. A key component was a device called an escapement, which rotated the gear train in the clock in regular ticks. By the 1400s, every big cathedral wanted a mechanical clock, and in making more and more impressive versions, artisans ornamented them with animated figures.[13] Initially most were simple processions of rigid figures, but extra animation was supplied to the "clock-Jack," a figure that hit the bell with a hammer to sound the hours. By the 1500s, more elaborate animated figures were included.[14] The clock in the Piazza San Marco, in Venice, featured not only two giant shepherds to strike the hours but also three Magi in procession. These bowed before the Virgin and Child and extended their gifts with one hand while removing their crowns with the other.[15]

Initially these mechanical clocks were powered by a cord wrapped around a pulley, but during the fifteenth century, mainsprings began to be used to store the energy needed to drive the clock. An advantage of a spring-driven system is that it can be miniaturized to make it portable, resulting in the development of pocket watches. This technology in turn brought a whole new approach to animated figures, which could be driven with much greater precision. They began to be known as *automata* because they could carry out complex movements without human intervention.

An early example of the new automata was the miniature monk attributed to the Italian Spanish clockmaker Juanelo Turriano and built in the 1560s. It is known as "the Clockwork Prayer." The Smithsonian Institution's National Museum of American History has an automaton that fits the description of this work, and it still functions. It takes the form of a monk somewhat over a foot tall, moving on wheels, with a turning rod to swing it around at intervals, and feet that rise and fall to make it look as if it is walking. One arm brings a rosary up and down, and the other beats the breast of the figure in the gesture that accompanies the Catholic "mea culpa" prayer. Meanwhile its mouth opens and shuts.[16]

If this automaton is the work of Juanelo Turriano, then it was commissioned by King Philip II of Spain after his son recovered from a serious head injury. Philip apparently believed that bringing in the relics of a fifteenth-century Spanish saint, Didacus of Alcalá, had made all the difference, and commissioned a clockwork model of Didacus. No record exists of whether Philip liked the automaton, but modern observers often find the museum automaton "creepy," possibly an example of the uncanny valley we cover in the next chapter. Though mechanically impressive, it falls into the then common pattern of courtly display and serves no purpose other than as a curiosity and an advertisement for the skills of its maker.

While spring-driven clockwork makes a free-standing figure and precise movement possible, there is a practical limit to the size and weight of such artifacts. A very large spring takes a great deal of force to wind, as well as being a much more difficult item to manufacture. It is also heavy, so that at some point the force needed to move a much larger automaton has to be traded off against the extra weight added by the bigger spring and the necessary gear trains. Spectacle, amusement, the gaining of prestige, and the provoking of wonder were the purpose of these automata rather than practical application. Indeed, it is hard to think of any practical application that would not have been better served by a human or by the slowly growing number of water- and wind-driven machines, starting with mills.

Just as today, however, so the construction of automata did provoke philosophical discussion about what it means to be human. If automata could be constructed to represent the human body, might this mean the human body was also a machine? If so, would this mean that humans themselves were machines? The French philosopher René Descartes, writing in the first half of the seventeenth century, when more and more elaborate automata were being constructed, offered an influential answer. He argued that human bodies and human minds were fundamentally different in nature, and while bodies were physical and took up real space, minds were neither physically nor spatially located. Descartes was clearly thinking of consciousness and self-awareness when he put forward this argument, hence his "cogito ergo sum"—I think, therefore I am.

This dualist view of body and mind suggests that a mechanism cannot itself count as *alive* unless it is imbued with some nonphysical component, a soul or a mind. This is why Pygmalion's statue had to be animated by Venus; also why the Jewish golem of medieval Prague had to have the word of God on its forehead to become more than the clay of which it was made. Later, when Frankenstein's monster was brought to life in Mary Shelley's influential novel, this role was played by electricity—seen as a mysterious force in the nineteenth century. In modern robots, we have artificial intelligence: some unexplained development in its capacity converts a machine into a self-aware entity like a human. The robot hardware is the body, and the software plays the role of the disembodied mind in this conception.[17]

Most people do not think of automata like Turriano's miniature monk as being robots. One argument against viewing them in this way might be that they could only do one mechanical thing: they had no flexibility, could not, as we would say now, be programmed. The work of the Swiss watchmaker Pierre Jaquet-Droz, in the second half of the eighteenth century, shows that this is not entirely true.

The Musée d'Art et d'Histoire in Neuchâtel, Switzerland, holds three automata crafted by Jaquet-Droz: the Musician, the Draughtsman, and the Writer. These are doll-size automata of great complexity. The Musician has 2,000 components, the Draughtsman 2,500, and the Writer 6,000. The Musician plays a miniature keyboard by actually pressing the keys rather than miming to a music box. Its head and eyes follow its fingers, and its chest simulates breathing. The Draughtsman draws four different images, periodically blowing on its pencil to remove dust. It foreshadows work in our own day on using a robot arm to draw portraits.[18]

The Writer is the most complex of the three automata and, from a modern perspective, the most interesting. It could write *any* forty-character string. It has a wheel allowing each of the forty characters to be preset, so it was indeed programmable, albeit via hardware settings, not via software. The Writer stands over two feet tall and writes with a goose quill, which it inks at intervals and shakes to prevent ink spilling. Like the Musician, its eyes follow the text it is writing. It has many of the capabilities of our own robots, other than size and being driven by clockwork. It can justify its tag as "the first robot," unless we want to demand that being able to sense its environment and choose actions in real time is a requirement. The Writer was prescripted.

The technological basis for modern-day robots was being laid in this period as the Industrial Revolution began. The idea of programming machines was not invented by Jaquet-Droz but was very much in the air, especially in the weaving of textiles. Demand for new textiles, damasks with complex patterns, had already led to looms that were directed by a paper tape.[19] The Jacquard loom, using punch cards to map a damask pattern onto loom actions, arrived a few decades after these automata. The first weaving factories also appeared in the 1780s, though mechanizing metalworking took longer.

Looking at the history of moving mechanical human figures—and we have only covered western Europe—makes it clear that we have been fascinated by the idea of human-like artifacts for a good 2,500 years. The idea that these things look like humans and have some human functionalities, but are not human, seems to have been a continuing preoccupation. Our own robots really only make up the last one hundred years of this long story.

1.1 This automaton is the work of another Swiss mechanician of the eighteenth century, Henri Maillardet. Known as the Draughtsman-Writer, it produces four drawings and three poems using a cam-based memory. Now in the Franklin Institute in Philadelphia.

We now jump about 150 years from Jaquet-Droz to 1921 and arrive in a very different world. Electricity provides the main power source; goods are mass-produced in factories with a growing level of automation. A world war between nation-states has just resulted in the mechanized slaughter of millions of young men. Europe no longer has slaves, and it no longer has serfs bound to the land, though their abolition is still in living memory. Parliamentary democracy with mass access to the franchise has replaced hierarchical societies dominated by aristocrats: hard laboring can no longer be carried out by fiat, and wage labor dominates.

This is also a post-Darwinian world in which the idea of the "survival of the fittest" has led to eugenics programs and so-called scientific racism in which certain populations are deemed less fit than others. Furthermore, Darwinism has removed the assumption that humans are God's chosen creatures and underlined that many species have gone extinct. Like a child discovering death for the first time, we understand that our whole species could go out of existence.

Why 1921? Well, if Pygmalion and Talos were foundational stories for the development of earlier robot-like entities, 1921 gave us the foundational story for the modern age of robotics. Karel Čapek's play *Rossum's Universal Robots*, or *R.U.R.*, gives us for the first time the word *robot*. The play is written in Czech, not in English, so *robot* is a translation of the Czech word *robota*, which actually means forced labor. The *roboti* take the place of the human slaves or serfs that are no longer available to carry out drudgery.

Ironically, given the way we now use the word *robot*, Čapek's robots were not machines made of metal. They were artificial creatures created by biological engineering, and therefore much closer to what we now call androids, descendants of Pygmalion's statue, rather than of Talos.

R.U.R. has all the elements of our current anxieties. Robots are produced by the thousand because they cut labor costs to a fifth of what they were; the robots are not paid. The world economy comes to depend on them, and then, inevitably, they revolt—as human slaves did so often—and wipe humans out except for one engineer. Most robots cannot reproduce, and the formula for their creation has been lost in the mayhem, but two advanced models fall in love, and at the end of the play, the woman robot is pregnant. In short, they take everyone's jobs, and then they supplant humans: the two themes that worry people now.

As we saw with Pygmalion's statue and Talos, while stories illuminate people's concerns, they may be some way from what can actually be built. One of our purposes here is to demonstrate how far we are in actual robot development from our own stories about robots. Many of these stories have been told in film; indeed, because most people have never interacted with a real robot, it is film above all that forms their impressions of what robots can do. Autonomous robots have not so far been used as film actors. The reasons include lack of capability, lack of reliability, and the need for someone on set with a deep technical understanding to reprogram the robot.

The most obvious way of getting a "robot" to carry out the right actions in a film is to put a person inside it. This idea goes back to the age of automata and the eighteenth century. In 1770 a wonderful new automaton went on display, known as the Turk, or sometimes the Mechanical Turk, or the Automaton Chess Player. The Turk wowed people with its chess abilities, able to play a strong game against human players as well as conduct "the knight's tour," moving the chess knight around the board, so that it visits each square just once. Several people suspected a hoax during the Turk's many European tours, during which the Turk played against both Benjamin Franklin and Napoleon. However, the Turk's secrets were not finally revealed until the 1850s, by which time it had fallen into disuse. It was in fact operated by a human hidden inside. Conjurer's cabinet trickery was used to hide the operator when the device's doors were opened, supposedly revealing its interior.

Early films solved their problem in exactly the same way: by putting a human inside a robot body. Sometimes this was obvious, like the Tin Man in *The Wizard of Oz* or his descendant C-3PO in *Star Wars*. Sometimes it was less obvious. The other famous *Star Wars* robot, R2-D2, required a very small actor because of its shape. Kenny Baker, who played the role, was only 112 centimeters tall. Robbie the Robot in the classic film *Forbidden Planet* (1956) also relies on an actor inside a robot costume.

The advent of motion capture technology meant that the human operator could be removed from inside the robot. This made designing interesting-looking robots easier. The film *Short Circuit* (1986) is typical of this period. Its robot, Johnny 5, was operated remotely by a human wearing a *telemetry suit*. The suit captured each movement of the actor and then sent a signal to the robot, which mirrored the movement. For close-ups, the robot was operated as a puppet, with metal rods attached to parts of it, manipulated by a puppeteer.

When film went digital, improved special effects technology made things even easier. Now the actor can directly play the robot, often in a motion capture suit, and then this movement is mapped onto a graphically generated robot model that replaces the actor in the digital scene. This means that the robot does not have to be a physical object at all and can carry out movements that would be impossible if executed in the real world. The bottom line is that though film forms most people's impressions of what robots can do, there is almost no relationship between what they see and real-world robots. However, as long as film fits in with people's prior expectations, its beguiling naturalism is liable to convince them that they are seeing the real thing.

Surely, though, the videos produced by researchers and robotics companies to demonstrate their own robots are more reliable? Such videos can be seen on the internet and have become increasingly popular. In this case, you are seeing real, physical robots, carrying out actions in the real world. However, these videos come with some important caveats. The first is that you are seeing only one instance of the robot's behavior; you do not know how many attempts it took to get this to happen. Robotics companies are clearly not going to supply their outtakes, showing where things went wrong. Most researchers do not do this, either, though you can view amusing online footage of outtakes from the US DARPA Robotics Challenge Finals of 2015, with legged robots falling over.[20]

A second problem with such videos is that clever editing can always suggest much more flexibility than a robot actually displays. The amount of work required to switch a robot from performing one activity to a different one is not visible. Finally, videos rarely show whether the robot is acting autonomously. Frequently a human operator is teleoperating the robot off camera. This practice is so widespread that researchers into human-robot interaction have a standard schema for it, known as "Wizard of Oz."[21] This refers to the scene in which Dorothy and her companions arrive in the Emerald City and are conducted into the presence of the Wizard, who is a terrifying, invisible presence with a booming voice. Then Dorothy's dog, Toto, dashes into a corner of the room, where there is a curtain, and pulls it aside to reveal a not very impressive middle-aged man pulling levers to create the wizard effect.

Creating robot capabilities is difficult and time-consuming. The Wizard of Oz approach allows researchers to investigate how the robot will interact with humans, using a remote operator, before putting all that effort into developing the robot's autonomous behavior and possibly getting the design wrong. Since the responses of participants are affected by whether they think the robot is autonomous, the remote operation is concealed during an experiment. Research ethics demand that participants are told that the robot was not working autonomously after the experiment finishes, but this is not always obvious in videos.

In other video cases, the robot was prescripted or is recreating a task it was trained for element by element. For example, one can take a robot around a building and allow it to form a map. The robot can then repeat the traversal on its own, as long as the route does not change and the environment remains reasonably static.[22] This is a very different thing from autonomous navigation. The interesting thing is how much viewers of films and videos want to believe in the capabilities of the robots they see. Film stories may not tell us much about real robots, but they certainly tell us plenty about ourselves and our anxieties.

Films about robots often present versions of the idea that the technology we create will turn against us. This is known as "the Frankenstein complex," after the monster created by

Dr. Frankenstein in Mary Shelley's novel. Like Pygmalion's statue, the monster is animated into life after being constructed by a human, though this time by a scientist rather than a sculptor. Where Pygmalion's statue is a beautiful female and becomes the sculptor's sexual partner, Frankenstein's monster is male and ugly and has no sexual partner, motivating its hatred for humans. It also has more than human strength, and though not a robot but an android, its murderous rampage and enmity toward its creator have been taken up by many robot stories since. The Frankenstein complex also informs other stories that do not involve robots, such as the computer HAL in *2001: A Space Odyssey*, one of a number of sapient supercomputers with similar behavior.

This idea runs so deep in our Western culture that we might think it is universal. Yet Japanese culture does not view either technology in general or robots in particular as potential Frankenstein's monsters.[23] So what makes us think like this, when other cultures do not? Perhaps our inheritance of the Greek concept of *hubris*—trying to outdo the gods—is involved. Hubris was inevitably followed by *nemesis* as the gods hit back. Moreover, the Old Testament god, as the Bible points out, is a jealous god and is seen as the origin of all life. Maybe we feel that building robots or androids usurps this power. Certainly the developers of scientific inquiry in medieval Christendom had to deal with accusations that they were guilty of the sin of pride. Our own cultural bias makes it all the more important that we think carefully about our fears and anxieties when it comes to robotics.

If we take a step back, the robots and androids in films often seem stereotypical. They are nearly always assigned a gender, though since they do not reproduce, sexual differentiation serves no actual purpose. Human characters in these films do not question this trope. "Male" robots or androids tend, like Talos, to be inhumanly strong; "females" may offer or exploit sexual attraction with male human characters, as does Ava, for example, in the film *Ex Machina* (2014). "Females" frequently have exaggerated secondary sexual characteristics—metallic busts and pinched waists—as seen in the front-cover illustrations of popular science fiction magazines.

It does not seem accidental either that the female T-X—Terminatrix—from *Terminator 3* sounds so much like "dominatrix." A refreshing, though isolated, counterexample to this sexualized portrayal of "female" robots is L3-37 in *Solo*, one of the *Star Wars* films. This robot is played by a woman but has an entirely machinelike appearance without any gender cues. Then we have EVE in the animation *WALL-E*; though this robot is portrayed as a flying white ovoid, clever animation nevertheless manages to suggest femininity, and the story produces a romance between EVE and WALL-E.

While gendering passes without comment, robot or android characters are often shown as lacking emotion. This is frequently highlighted as their distinguishing feature, especially in the case of androids, whose appearance is as human as the human characters. The character Data in the *Star Trek* franchise is a benign example. An earlier *Star Trek* character without emotion, Spock, was actually portrayed as an alien rather than an android or a robot. This trait stands Descartes on his head: what makes us human is not our ability to reason but our ability to emote. However, we will see in chapter 10 that good reasons exist to give robots a model of emotion as part of their overall model of intelligent behavior.

Robots and androids may involve stereotyping in a deeper sense. *R.U.R.* plays on the idea of an artificial species, as we say in our post-Darwin age, with entities that look human but are fundamentally not. However, we have always made these distinctions. Historically, the other would have been foreign invaders, or going back far enough, a different tribe, or even people from the next village.

Stereotyping other humans as not properly human like us is a continuing thread in human behavior, as populist politicians know well. In the present, we see how easy it can be to mobilize populations against immigration, with a reflex fear that "they" are going to take things that "we" need. Psychologists would describe this as in-group and out-group thinking, and if some of our stories recast "them" as androids or robots, the only difference is that "we" have made them. In talking about robots as "them," we are often talking about what we think being human means; the discussion really is all about us. And we can be our own worst nightmares.

2

APPEARANCE: WILL THEY LOOK LIKE US?

In January 1979, a twenty-five-year-old Ford factory worker named Robert Williams said goodbye to his wife and three children and set off for what he thought would be an ordinary working day at the Ford factory in Michigan. This factory used a monster five-story machine to move heavy molded car parts on and off racks, and on each story a metal cart with a robot arm could lift, store, or retrieve these car parts. But there was a problem. Information from the machine was reporting the wrong number of parts up on the rack, and Robert was asked to climb up to find out why. While he was investigating, a robot arm on one of the carts moved around and struck him in the head. The blow to Robert's skull killed him instantly, and in this unfortunate accident, Robert Williams went down in history as the first person on record to be killed by a robot.[1]

This was certainly a tragic health and safety fail, but many who work in the field of robotics would argue this machine was not really a robot. If you ask a robotics researcher for a snappy definition of *robot*, they might well come out with the phrase "physically embodied agent." Let's unpack this definition.

The word *physically* means a robot takes up physical space in the same world as us, so a computer game character would not count. It also means a robot is subject to real-world physics. It takes energy to move a robot. A robot is subject to inertia, momentum, friction, and other physical quantities. As roboticists will tell you, sometimes ruefully, robots are also subject to physical decay: they can rust, suffer dents and breakages, and their internal soldered contacts can work loose.

What about the word *embodied*? That seems straightforward. It means the robot has a body, unlike pieces of software on the internet, for instance, though these may be referred to as *bots*. A body has a physical extent, the boundary of which defines how the robot interacts with the world, from essential capabilities like mobility to sophisticated social interactions such as facial expressions.

Using the word *bodies* produces an assumption we are not always aware of. Living things have bodies. Fridges, mobile phones, and most machinery may exist in the physical world, but we do not think of such objects as having bodies (though we do use the term for cars, the machinery in which we ride).

The word *agent* implies action. An agent has agency, can act independently, and is autonomous. We normally assume that an agent is able to choose from a range of actions and do the right thing in a specific situation. This requires the capability to sense what the situation is, assess what the right action should be, and carry out that action. While *physical* and *embodied* seem to be attributes you either do or do not have, *agency* is a matter of degree.

Does a thermostat that switches a heater on or off in response to the room temperature count as an agent? Believe it or not, this is a tricky question, and not everyone in robotics agrees on the answer.[2] After all, the thermostat is assessing the current situation—the room temperature. It is selecting whether to switch the heat on or off or leave it as it is, and it can execute one of those actions.

Many of us might feel that deciding whether to toggle a switch is not enough for real agency, and responding to temperature measurements does not require anything we would call intelligence. But would a really sophisticated thermostat, using knowledge of the seasons, the number of people in the house, the current state of the fuel bill, and individual preferences for temperature, count as a robot? Most people would say not, since an intuitive addition to the earlier definition is that the physically embodied agent should be able to move at least some part of its body, if not all of it. Movement is indeed very important in robotics, and we will come back to it in the next chapter. But first let's consider the design of the physical body—how our future robots will look.

Designers of artifacts from buildings to kettles distinguish between form and function. A good design marries the two, but usually one of these—either the form or the function—takes priority, depending on the environment in which the artifact operates.

Real-world robots were initially deployed as an extension of factory automation. A factory is an environment where efficient functioning is the most important aspect of design. Industrial robots pick objects up, manipulate them, and put them down with great precision and repeatability, and they are usually shaped as what we would call an arm.

Their design reflects the functional requirements of their tasks. Some are roughly based on the structure of a human arm, with a joint at the top (shoulder), halfway down (elbow), and near the working end (wrist). Beyond the equivalent of their wrists, they are equipped with end effectors. These end effectors could be grippers, loosely based on fingers; or as is more often the case, they will be specialized tools. Other designs for these mechanisms look less like a human arm but still involve joints because they allow the robot to successfully manipulate objects.

Industrial robot manipulators are heavy pieces of machinery, made of metal and driven by motors. They need to be constructed with enough solidity to prevent the arm from

2.1 A modern industrial robot with two arms that can cooperate. This one has sensors to stop it from hitting anyone.

wobbling or flexing as it moves. They are not the kind of thing a human should stand too close to. Until quite recently, industrial arms had no sensors that would allow them to notice someone in their path. Like any heavy moving object, robot arms also possess a great deal of momentum, and this is what killed Robert Williams. Physical bodies can do physical damage.

This is why most automated factories are organized to keep robots and human workers well apart. Reorganizing a factory for robot automation is one of the major costs of introducing industrial robots, much more expensive than the robots themselves. Of course, accidents can still occur, and Williams was not the last person to be injured or killed. Maintenance workers testing the arms are at greatest risk. The problem is not that the robot intends to do harm but that it does not have intentions at all. It operates mechanically, whether or not any humans are nearby. This is no different from the harm any other piece of factory machinery can do to a human who gets in its way. The dreadful history of factory industrial injuries bears witness to what can happen when heavy machines come into contact with soft human bodies.

The reason many roboticists don't view most industrial robots as real robots, by their "physically embodied agent" definition, is that although industrial robots do move, they reliably repeat the same sequence of operations indefinitely. These actions are preprogrammed into them, which means they are not making decisions in any real sense of the word. Many of them have even less agency than the thermostat. At least the thermostat is making its simple decisions in response to the surrounding temperature changes in real time.

In other systems referred to as robots, the same sort of thing is true. Planetary rovers are sent sequences of instructions, but if one of the instructions fails to work as intended, the fallback is to stop and wait for a new sequence. Bomb disposal robots are usually teleoperated, meaning that a remote human operator decides what actions the robot should carry out. In both cases, the environment is hazardous, and the consequences of errors are serious, so that designers are reluctant to introduce autonomous decision-making. The fear is that in an environment not tailored to the robot, its autonomous decisions might not be correct. We saw in chapter 1 that even filmmakers worry about this problem and so far have never used autonomous robots as actors.

Back in 1979, the psychologist James Gibson discussed a new and important design consideration he referred to as *affordance*.[3] An object's affordance is the way its function is communicated through its design; in other words, how an object makes it obvious to a human what that human can do with the object—what functionality it affords. A round doorknob is for turning, and a door lever is for pressing down. Affordance says that where humans interact with objects, their form should reflect their function in a humanly intuitive way. A robot arm is not only functionally designed to lift objects but also designed to make it obvious to us that lifting is what it is supposed to do.

The same applies to *social* robots. Social robots are not isolated in an industrial environment tailored specifically to their capabilities. They are designed to work in ordinary human environments alongside people carrying out their own activities.

Researchers discovered early on that most people see robots as social agents, with their own goals and intentions, even though they are in fact pieces of machinery.[4] In an experiment conducted in the United States, with participants working on a task jointly with a robot, one participant asked the robot, "Would you like me to put this item into the bucket?" The robot turned and drove away, having failed to register the question. This left the participant perceiving the robot as intentionally rude.[5]

Because we humans are inherently social beings, we ascribe social purpose to all kinds of things that in reality do not have any. Trees and rivers once had spirits, specific animals were patrons of clans and tribes, and you may see a notice above the photocopier, "Never let this machine know you are in a hurry."

This even goes for pictures. We associate eyes so strongly with being watched that nothing more elaborate than static images of eyes can change people's behavior. When a team of university researchers alternately placed pictures of flowers and eyes each week over the honesty box for staff coffee contributions, the weeks with pictures of eyes produced significantly higher contributions. In a similar experiment, another group of researchers found that displaying pictures of eyes increased the likelihood of people clearing up their litter.[6]

If you give a robot eyes, even if these eyes actually provide no information to the robot, then people will act as if they work like human eyes. As a result, designers often add a "blink" function for robot eyes. The robot does not have to blink as a human does to keep its eyes moist—rather the opposite, as blinking uses motors and wears them out much too quickly. But unblinking eyes upset people because they look as if the robot is staring at them, which is uncomfortable, rude, or even aggressive in interactions between humans.

This is why robots designed to work with us not only have a functional affordance that makes visible what they can do, but also have what could be called a social affordance. A robot's body keys into what people expect of it—how well they think it is doing, whether they like it or fear it—and those attitudes are fundamental to acceptance and smooth interaction.

While a factory is a standardized environment, our everyday social environments are far more varied, and the social affordance of a robot body depends on where it will operate. The social affordances of a robot designed to carry equipment around a hospital might not work for a robot designed to carry objects in someone's home. Human expectations are related to social roles, and social roles are usually attached to specific environments. In the case of the hospital, the robot acts like a porter. In the domestic setting, it may be acting like a companion or even a carer.

What tasks the robot is supposed to perform also influence how people expect it to look. A Japanese experiment in which robots were put into the homes of elderly people for six weeks found that they preferred fluffy-looking robots for companionship functions but a more mechanical look for a robot that reminded them to take their medication.[7]

One thing we can do with robot bodies that we cannot currently do with human bodies is to reconfigure them. Even industrial robots may have a variety of tools attached to the working end of their arm: grippers if they are to pick things up, a screwdriver or a welding tool for assembly. We can imagine a modular robot that would add or subtract body elements depending on what tasks it had to carry out, perhaps adding a tray for a carrying task, a polishing device for cleaning, or even an expressive head for social interaction.

Swedish researchers investigated an interesting variant on this idea using items of clothing on a small toy dinosaur robot, the Pleo.[8] Users could personalize their robot's behavior—not just its appearance—by adding or subtracting such items. Imagine a domestic robot in a house where people support different football teams. Adding the team scarf could reconfigure the robot to act as a fellow supporter, accessing relevant information about the team, praising its performance, or commiserating over matches the team had lost.

This kind of personalization may be necessary if robots are to be accepted into our everyday environments, if only because preferences vary so much. A group of researchers in the United Kingdom ran experiments to see what people felt about the appearance and also the height of a robot.[9] They tried a shorter or taller robot, each presented either with a schematic face on top or with just a camera. The researchers found that their participants preferred either the taller robot with a face or the shorter one without a face. People also felt

2.2 The Pleo was a small dinosaur robot that researchers were able to develop new behaviors for.

that all the taller robots overall were more humanlike and conscientious than the short robots.

So it seems that for robots, size does matter to people who interact with them. Robot manipulators are often larger than an adult human, never mind a human arm, and their metallic bulk and machinelike appearance give them an intimidating aspect. Robots designed to appear as pets or toys are much smaller than a human, so they can be easily handled. They rarely operate on the floor, unless as children's toys, because they could form a trip hazard. They are more often designed to work on tabletops. If humanoid in appearance, then they suggest dolls, babies or children, and otherwise the sort of animals—real or fantasy—with which humans enjoy interacting.

On the other hand, robot vacuum cleaners, though also small, are usually circular and relatively featureless. They are designed to be socially invisible, and operating quietly is a significant functional requirement. The idea is that people will see them purely as machines, though in fact there are examples of owners customizing their appearance, as the Pleo owners did their dinosaurs. Possibly we also see them as the sort of animal we disregard because engrossed in some purpose of its own, like bees or ants. Going down a level in size to what are known as nanobots—robots that might one day be able to work inside a human body—we feel this analogy even more strongly. Researchers in the United States recently experimented with what they called origami robots.[10] Shaped like small cubes, they drive themselves into different exoskeletons, giving them a range of capabilities—as a tiny boat, with miniature legs, or a wheel.

Some robots are not designed to move around in physical space at all, though they usually have heads or faces that can move. These are desktop robots, embodied versions of the conversational interfaces such as Siri, Alexa, and Google Home. Designers seem to have little consensus about how a desktop robot should look.

Some are inspired by animals: a rabbit, a frog, or a cartoonlike cat. Others draw on approaches from film animation and look like a moving desk lamp, with a head/face where the lampshade would be. At least one desktop robot has a graphically animated face projected inside a glass head shaped like a human's—though not everyone is comfortable interacting with a humanlike head that has no body attached to it. These designs offer very different social affordances, reflecting an attempt to introduce a new social role without obvious precedents.

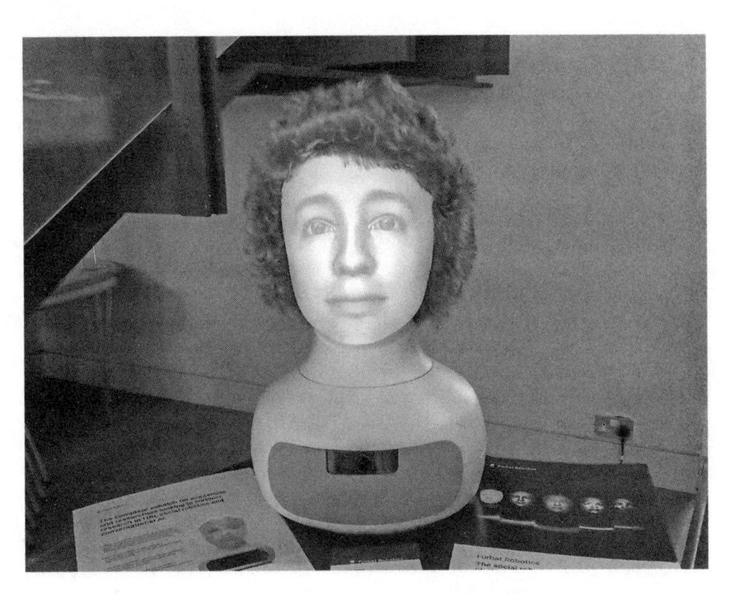

2.3 The Fur Hat robot is a glass model of a head with an animated face projected from inside. It is wearing a wig in this picture.

On the other hand, robots designed to move around indoor environments and interact with adult humans need enough size and bulk to be treated seriously as interaction partners without being so bulky that they intimidate. The established rule of thumb is to make them the height of a seated adult, equivalent in some sense to a human in a wheelchair. However, if they include a robot manipulator, the more-than-human-size shoulder they require can give the impression of menace, as some found with the recent Baxter robot.

So this is where function comes back into the picture. Designing a robot arm that can manipulate household objects but at the same time has proportions more like a human arm is beyond the current state of the art. Therefore, if the robot is operating in a social environment, equipping

2.4 The Baxter robot.

it with well-functioning arms raises serious health and safety issues. How can one prevent people being injured by standing too close to the robot in an environment they both share? Remember that the sheer bulk of industrial robot arms is one reason that they can be dangerous. As we will see in a later chapter, we can reduce the chances of injury by including sensors in the design, causing a robot arm to stop before it hits anything. But this does not solve the problem of the intimidating appearance, making people reluctant to share their environment with such a device.

The same interaction between form and function affects whether a robot has legs or wheels. Since human environments are designed for two-legged beings—just ask a wheelchair user—it may seem obvious that robots should have legs, too. Unfortunately, the reason why children take some time to learn to walk is because two-legged locomotion is inherently unstable. Walking is a kind of controlled falling, as we will see in the next chapter. A chest-high robot made of metal is heavy, and if it falls on top of somebody, they will certainly feel it. Four-legged robots are more stable but produce a more horizontal body more difficult to turn, as well as having the social affordance of an animal. Wheels are much safer, but a wheeled robot needs a flat floor and struggles with steps. It also suggests *machine* to a human observer, as do tanklike tracks.

Just as functional affordance tells us what a robot can physically do, so its social affordance affects how we expect it to interact. Again, both the physical and the social are involved. A robot without a face cannot make facial expressions; a robot without arms cannot make gestures. Pressing buttons on the body of a robot that does not have a face feels very different from doing the same thing on a robot that does. If a robot looks humanoid, we may well expect it to communicate through speech.

Ducking high expectations is why many robots are not humanoid, modeled more on machinery or on animals. Dogs, birds, and snakes all figure. If a robot looks like an animal, a range of beeps and chirrups might be acceptable, and we might focus on the way it wiggles its body as communication. If it looks like a machine, then communication via colored lights and text displayed on a screen may seem relatively natural. If it seems like a social insect, we might not expect it to communicate at all but would watch its behavior to work out what it was up to.

Our expectations of robots are strongly colored by fiction, as we saw in chapter 1, especially by robots in films. After all, most people have still never encountered a robot in the real world. If you ask people to draw a picture of a robot, they usually make it look humanoid—like the Tin Man in *The Wizard of Oz* or C-3PO in *Star Wars*. They also tend to give the robot a gender—by default, male. "Tin Man" was a common description of what we now call a robot before the term *robot* appeared.

So should we make our robots look as human as possible, maybe like the androids of films or TV series? Should we go along with the idea of robot gender and provide visual and behavioral cues we recognize as "male" or "female" when we see them in humans? Do we provide bumps on the torso to suggest breasts, or a metal chest ridged to suggest a muscled male? Do we make faces that look like a stereotypical woman, down to lips colored to suggest the use of lipstick? Since we know that people expect humanoid robots to use speech, should we give them a male voice or a female voice? Or do we go for a machinelike voice?

Some Japanese researchers have opted for the as-humanlike-as-possible approach. Their robots have latex skin, glass eyes, pink lips, and synthetic hair on their heads. In most cases, they are designed to look like young women, an approach that some might see as worrying, a reflection of cultural assumptions about the role of women in their society. It is a curious echo of Pygmalion's statue.

It turns out that there is a serious problem with high-fidelity humanlike robots. Ironically, this was identified by another Japanese researcher, Masahiro Mori, back in 1981, well before such robots could be built.[11] Mori reasoned that as a robot looks more human, we like it better because it looks more like us—that is, until it looks almost entirely like us. At this point, he suggested, a precipitous drop in liking will occur, right down to people feeling extremely negative indeed.

Why? His answer was that the increasingly human appearance triggers stronger and stronger expectations about being able to apply human assumptions to the robot—until a point when it is perceived as almost, but not quite, right. And that not-quite-rightness makes us feel that the robot is uncanny, weird. Mori's diagram, with its sharp drop in liking, marks what researchers call the uncanny valley. We could equally call it the zombie zone, since zombies also look human in the wrong kind of way.

This uncanny valley reaction can certainly be triggered by some of the Japanese humanlike robots.[12] It is worsened when we perceive a mismatch between the humanlike appearance and the way the robot moves. Unfortunately, robots that are driven by motors are inevitably somewhat jerky—and seeing jerkiness in something that looks almost human really upsets people. It is no accident that this is exactly how zombies move.

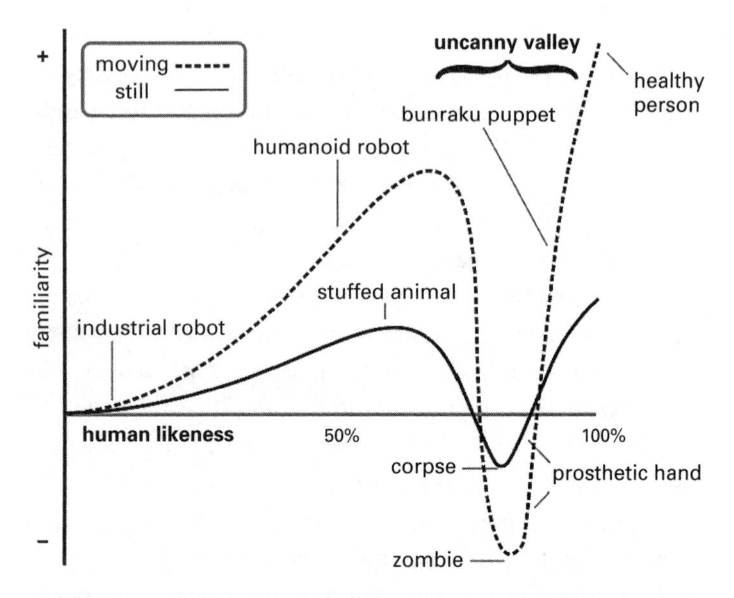

2.5 This graph shows Mori's hypothesis that we like things that look like us more until they are "nearly right," when we can suddenly feel very negative. Mori believed that the effect was even greater for movement than appearance. From Wikimedia Commons, accessed December 3, 2020, https://commons.wikimedia.org/wiki/File:Mori_Uncanny_Valley.svg. CC BY-SA 3.0.

Moreover, people interacting with a very human-looking robot inevitably focus on the robot's face, as they would with a fellow human. However, human faces are highly complex, driven by more than forty groups of muscles. Getting a humanlike robot face to move autonomously in the right way during speech, or for emotional expressions, is well beyond the current state of the art. Yet people interacting with such robots have strong assumptions that this behavior will be there, and strongly negative reactions if it is not.

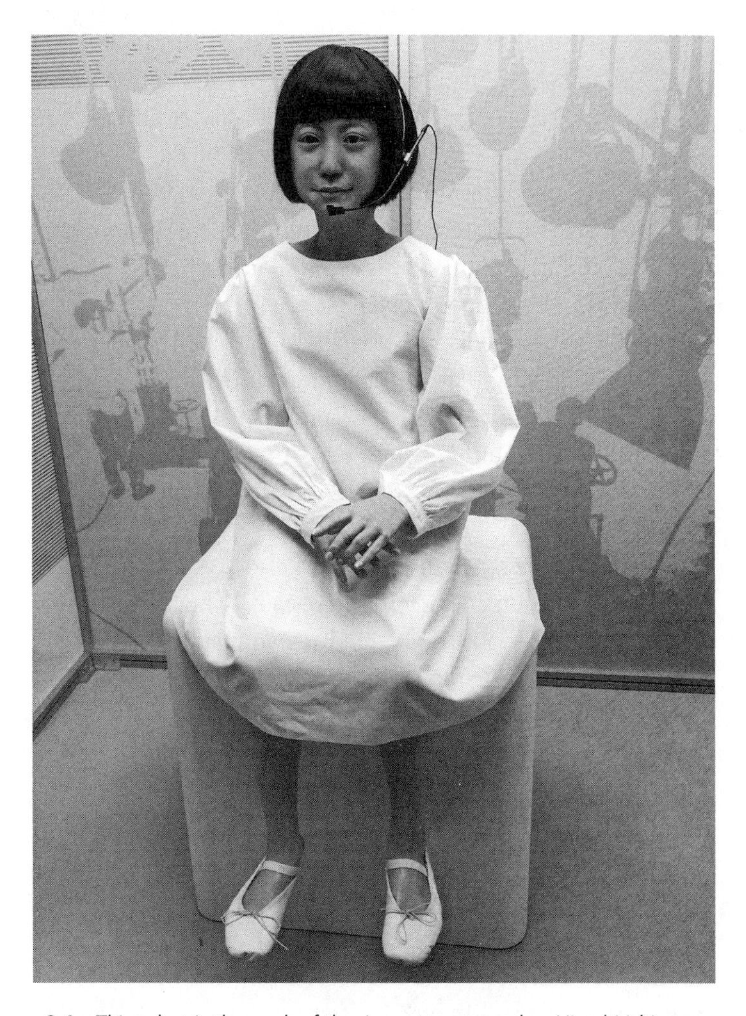

2.6 This robot is the work of the Japanese researcher Hiroshi Ishiguro, who specializes in very humanlike designs, mostly of young women, but including one based on himself. Many feel these provoke an uncanny valley reaction.

Lip synchronization—the way the mouth moves in step with the sounds we hear during speech—is a good example. Anyone who has watched a television broadcast in which the sound lags slightly behind the movement of the reporter's mouth will know how sensitive we are to whether this works correctly. The discomfort of dubbed films, where the sound is not related to the actual words being spoken by the actors, is similar. Errors in the face movement of a high-fidelity humanlike robot are immediately obvious to a human interaction partner, so much so that such robots are usually teleoperated, using real-time motion capture. Sensors pick up movement on a human operator's face and send it to the robot face.

Yet nobody expects low-fidelity cartoon characters to lip-synch so accurately. This may be obvious with Donald Duck—who has a beak, not a mouth—but it is also true of low-fidelity humanlike cartoon characters. These usually have only three or four different mouth shapes in their animation.

So strong is the effect of human expectations when we interact with a robot that some researchers have produced designs specifically intended to reduce those expectations. This idea was behind the experimental Kismet robot developed at MIT.[13] It is designed to look and behave as a young mechanical creature, with big eyes, a rubbery red mouth, and large pink ears.

If trying to make robots look as human as possible is a bad idea, then what should we make them look like? The discussion earlier in this chapter suggests an answer. We should give robots the physical and social affordances needed for the environment in which they are to work, the tasks they are to carry out, and the interaction style we want with the humans around them. This may seem a disappointingly vague answer, but it is a concrete version of "it depends."

Rather than aiming for the naturalism of robots in film and risking uncanny valley responses, we might aim for what researchers—and animators—call *believability*.[14] Think about some of the classics of animation: Mickey Mouse, the Ugly Duckling, Tom and Jerry. None of these characters are depicted naturalistically; it would be a shock to find a mouse in your kitchen that looked like Mickey. Yet we find them believable as characters, though we know they are merely

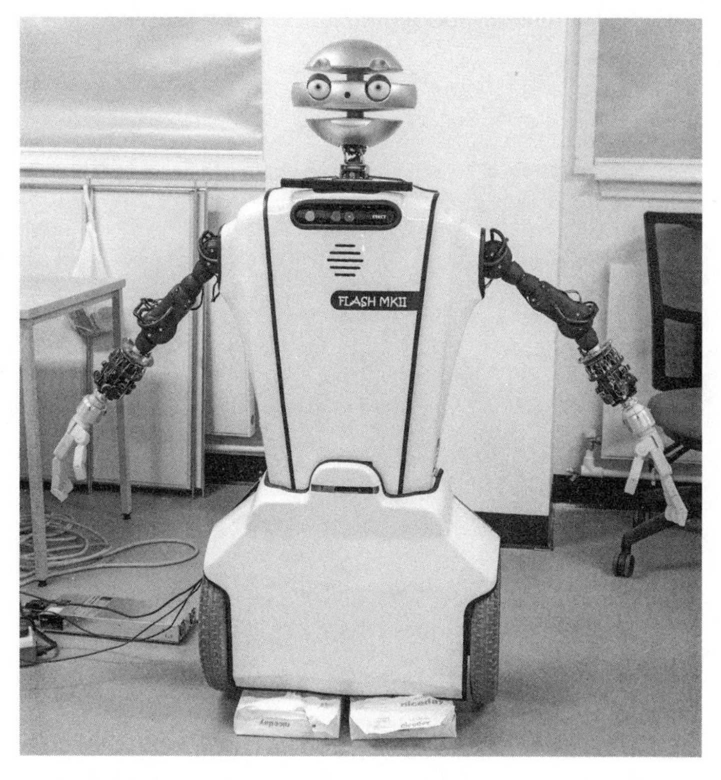

2.7 This Flash robot—known as Alyx—is an example of a humanoid design related more to cartoons than naturalism.

collections of pixels on a screen. When the ugly duckling is rejected as "not a proper duckling," the animator makes its body language so evocative that it is hard not to feel a deep pang of sorrow. In other words, we act as if these characters had their own internal states the way we do, and we could stand in their shoes and feel what they feel.

Some designers of social robots have taken this to heart and are more inspired by cartoon characters than naturalism. For them, WALL-E is a better template than the Japanese humanlike robot or the Terminator.

3

MOVEMENT: WILL THEY LIVE WITH US?

In 1507, King James IV of Scotland and his court watched as an Italian alchemist, known as John Damian, launched himself off the top of Stirling Castle. He was wearing wings he had designed using the ideas of Leonardo da Vinci. Unsurprisingly, he crashed and was lucky to come away with only a broken thigh. The interesting thing was that he blamed his failure on using the wrong kind of feathers for his wings. He had used chickens' feathers rather than eagles', and everyone knows chickens are very poor at flight.

It seems the king dissuaded John Damian from testing this assumption with another attempt. We may laugh at the explanation, but it underlines the amount of scientific analysis and engineering know-how needed for the Wright brothers to successfully take to the air nearly four hundred years later—without any feathers, in a machine that did not flap its wings.

So what has this to do with robot movement? The answer is that we cannot assume that robot capabilities must work like human capabilities. Engineering solutions may not be the same as biological ones. It is true that two-legged movement seems desirable for a humanoid robot. It meets people's expectations, and our own environments are designed with this kind of locomotion in mind. But if we analyze how humans walk, we find it is not easy to replicate on a robot.

One reason is the complexity of the human skeleton. Engineers producing software to control a structure use a concept called *degrees of freedom*. This is the number of independent movements a structure can carry out. If we think of something like a drone, flying in 3D space, then it has six degrees of freedom. Three of these allow it to move in a straight line up and down, backward and forward, and side to side. The other three degrees of freedom allow it rotate around each of these axes. This means that its control system needs six numbers for each movement to be able to relocate it to an arbitrary point.

A simple view of the human leg suggests that it also has six degrees of freedom: three from the ball-and-socket joint at the hip; one at the knee, which moves like a hinge; and two at the ankle, which can move up and down in relation to the knee or side to side.

In fact, human joints are made up of many more components, especially the ankle and foot, which contain more than twenty-six bones and thirty-three joints. Even the knee is not a simple hinge, since as it flexes it also automatically rotates slightly. This means that each leg has many more than six degrees of freedom. Each of those degrees of freedom must be controlled, so an accurate model would be calculating far more than six numbers for each movement.

That is not the only problem. With six degrees of freedom, a structure needs just one set of numbers to move from one point to another in 3D space. Add just one more degree of freedom, and the structure gains what engineers call *redundancy*. This means that now more than one set of numbers will produce the desired movement. Each added degree of freedom adds more redundancy, and the control system has to choose between the various options, as well as generating a bigger set of numbers. In some ways, this is helpful, since it means movement can be planned in more than one way. However, adding many extra degrees of freedom is so computationally demanding that engineers use biped models with far fewer degrees of freedom than a human has.

The next issue is the role of sensors in controlling human motor actions. If you ask how many senses humans have, most people reply "five": sight, hearing, touch, smell, taste. These are certainly important in our interaction with the outside world. However, we also have a number of internal senses. Two of these are vital to our ability to walk: proprioception and equilibrioception.

Proprioceptors in our body, especially at our joints, tell us how fast a limb is moving and in what direction, whether it is bearing a load, and when we cannot move it farther in a given direction. Proprioception is the kinesthetic sense that allows us to touch our noses with our eyes shut and tells us whether we are standing upright or upside down in a pitch-dark room. The brain uses information from its network of these sensors to maintain a map of where our body is in space and what it is doing. The same system allows us to execute a motor movement with a good idea of what the limb will do as a result. Exactly how the brain does this, we do not yet know, though we can observe babies developing motor control over their first few years, and we know which sections of the brain are involved.

When we walk, we feel our leg going forward and the force with which our foot meets the ground. This allows us to walk without watching where our feet are much of the time. Our walking motion can be described as a double pendulum. We swing a leg forward from the hip in a pendulum motion. Then the heel strikes the ground, and the foot rolls forward, bringing the body over the top of the leg as an inverted pendulum. There is plenty of scope for unbalancing and falling in this motion. This is why equilibrioception—or our sense of balance—is so important.

This sense is provided by our vestibular system, which sits inside our ears, behind the eardrum. Movement involves both translation—change of position—and rotation. The vestibular system has components for each. Translation is handled by organs called *otoliths*, which detect linear acceleration both horizontally and vertically. Rotation in a 3D world needs three structures, one for each type of rotation: nodding, turning the head, and tipping it sideways toward the shoulder. Each ear has three semicircular canals containing liquid set at right angles to one another. The movement of the liquid tells us whether we are in balance or not, working like a spirit level. Because the canals are in pairs, we can distinguish rotation right to left from its opposite.

As well as allowing us to walk without falling over, at least most of the time, the vestibular system stabilizes our visual system when we are moving. Otherwise our own motion would make the world move in our visual field. Anyone who has seen the bouncing camera image from a robot in motion will appreciate how useful this is.

The proprioceptive and vestibular systems work together to integrate knowledge of position and acceleration, again in a way for which we do not yet have a detailed account. As with many other human capabilities, engineering is up against too many areas of uncertainty about how our bodies and brains function to reproduce the human proprioceptive and vestibular systems and their interaction in a robot.

The differences between human and robot motion go well beyond the issue of how far internal sensors are involved. A rigid metal structure driven by motors, as most robots are, is very different from a living body driven by muscles. One major difference is its degree of *compliance*. Compliance is the opposite of stiffness: our feet bend when they hit the ground; our muscles can be squeezed and also change shape when they are working. Compliance is extremely useful in dealing with an irregular environment and is increasingly added into robot systems. However, doing this makes them more complex to build.

All these differences make it a bad idea to copy a human walk onto a robot using the approach of films and computer games. They animate their characters using motion capture technology. A human actor wears a motion capture suit that has markers on all its joints. Then the actor's movements are captured by a camera and mapped onto a sticklike graphical model of a skeleton. This skeleton can be "clothed" with a graphical body, and the resulting virtual actor will perform the natural-looking movement of the human actor.

Mapping human walking onto a physical robot is not nearly as easy. Graphical characters do not have to deal with real-world physics. The differences between the human body and the robot body mean we need to adjust the human walk by using what is called a *transfer function* as part of the mapping process. Working out what this should look like is extremely tricky. Moreover, the robot has to respond in real time to the world around it, which may not be similar to a motion capture studio. Even indoors on a flat surface, the robot will have to avoid static obstacles, like furniture, as well as dynamic ones, like people moving around it. This means stitching together different pieces of motion capture in real time, with any errors likely to result in the robot toppling over.

The simplest possible walking motion would calculate where the robot foot should go next. To make it easy, let's assume the robot is walking on a flat floor with a constant stride length, so that this position is straightforward to calculate. The branch of mechanics that deals with position is *kinematics*, and it can tell us for given joint angles of the robot's leg—at hip, knee, and ankle—where the foot will end up. Of course, that is not quite the problem we have to solve here. We already know where the robot foot should go, and we need to work out what joint angles are required to get it there. This is called *inverse kinematics*, since we are working the calculation backward.

Inverse kinematics turns out to be mathematically harder than forward kinematics because it involves inverting a matrix, a table of numbers. Inversion is a mathematical operation a bit like creating a fraction from an integer, but much more complex because it uses a whole table. Not all matrices can be inverted analytically. In some cases, the process passes through a divide-by-zero operation, and since any division by zero gives an infinite answer, the output to the motors controlling the joints becomes undefined: not what we want in a heavy metal structure. Various techniques now exist for avoiding this situation, so we will assume one is in use and the joint angles needed are successfully calculated.

The next step is to calculate the forces required to move the limbs so that the robot's joints arrive at the desired angles. This branch of mechanics is called *dynamics*. We need to know the mass of the limb and the overall speed at which it should move. Then we need to calculate the acceleration profile. We need some force to accelerate the limb and some force to decelerate it, too, since otherwise its momentum might cause it to overshoot or result in a robot foot hitting the ground rather hard. We have to decide whether to move the joints one at a time, simultaneously, or something in between.

We also need to think about the robot's center of gravity, since when this goes outside the base of support of the two legs, the robot could fall over. This is bound to happen because lifting a leg off the ground gives us only one supporting leg, which is why walking is a kind of controlled falling. We can tilt the body of the robot forward as the leg moves, and if we calculate this correctly, the dynamics of the movement will keep the robot upright.

This simple approach is an example of what engineers call *open-loop control*, since it works by calculating the motor actions needed, sending the control signals, and going on to the next motor action. If it sounds fallible, then that is because it is. The human system is equivalent to *closed-loop control*, since our internal sensors feed back information about what is happening and allow the system to change its behavior as a result.

We can add accelerometers to measure the acceleration of the leg segments when they move. Then, if the acceleration is not the one anticipated, we could take some remedial action. Knowing what action to take depends on exactly what has gone wrong—which is why dealing with errors is tricky. We can also add tilt sensors to do some of the work of the human balance system. Again we have to decide exactly what to do if the sensors give unexpected values, especially if they indicate the robot is starting to unbalance. Closed-loop control is inherently more complex than open-loop control, since the extra information has to be added into whatever the control system is trying to do next.

Early walking robots took the simple approach just described, but this was never very successful. It involved a highly inefficient use of energy, and as we will see later, energy cost is a big issue for moving robots. It also made walking a slow process. This worsened the balance problems: the longer it takes to get the robot's center of gravity back over its advancing foot, the more time there is to lose balance altogether. Slow walking is unsteady walking. These problems made researchers look again at their whole approach. They found two rather different sources of inspiration: Wobblebots and cockroaches.

Cockroaches turn out to be the fastest thing on six legs we know of. What intrigued engineers was that we also know quite a bit about cockroaches' rather simple brains. It was clear that they were moving their legs much faster than their brains could carry out the necessary calculations. This had to mean that the legs were being controlled locally, and not by the brain.[1] In a fairly repulsive experiment, scientists removed a cat's cortex and put the cat on a treadmill. Not only could the cat still walk, but it could change its walking speed as the treadmill moved faster or slower.

Biologists concluded that the legs of most animals are controlled by what they call a central pattern generator (CPG), a network of neurons close to the legs—usually in the base of the spinal cord for mammals—that produce a rhythmic output. This is why you can "get into your stride" as this neural circuitry patterns your walking. A set of CPGs has another interesting feature. Physicists have long known that if you set two pendulums going on a mantelpiece with different swings, they will gradually converge and swing together, a process known as *entrainment*. This means the cockroach can move three of each six legs in the same rhythm even though each leg has its own pattern generator.

Although a CPG does not need sensor information to work, it can be pushed into a different rhythm by a sensor, allowing an animal to adjust its gait to different terrains. It can also be given a "go faster" instruction from the brain to allow the animal to begin to run using a different rhythm from its walking. Because of the entrainment effect, a change in the movement pattern of one leg affects all the others, so that these gait changes emerge without having to be directly programmed.

Finding out how to control multiple legs quickly and effectively made them an attractive prospect for robots. A six-legged robot will never overbalance if it has a trio of legs on the ground at once in a triangle formation. Even a four-legged robot is much more stable than a biped, especially on uneven terrain. Multiple legs seem a good idea for many robot applications, for example, as a donkey-like carrier on rough ground, trialed for military use with infantry.[2] Search and rescue in disasters is another area where a multilegged robot seems the best choice.

One research group took the cockroach idea even further. Inspired by the use of dolphins with harnesses for undersea tracking, the scientists used real cockroaches, adding control mechanisms to them via a tiny backpack plus a camera.[3] The idea was that the cockroaches could be directed using electrical impulses and used to investigate small spaces with their superior insect mobility.

However, there are also many robot applications for which multiple legs are not the answer. This is partly because of the high power requirements (discussed later), but also because the social affordance (discussed in chap. 2) is certainly not that of a person. A cockroach-like, or worse still spiderlike, robot is not something everyone is comfortable with. Even a doglike robot produces different human interaction behavior from a more humanoid form. So the success of multileg robots has not stopped research into bipeds.

"But where do Wobblebots come into this?" you may be asking. If you have never seen one, a Wobblebot is a small plastic figure that can be placed on a slope. Its weight will cause it to walk down the slope without any battery or motor to drive it, or any CPG, for that matter. Once it gets started, its momentum just keeps it going. This is known as *passive walking*. Rather than a complex calculation about where to put its foot down, the forces acting on the Wobblebot—its dynamics—make it walk.

We cannot arrange for robots to walk downhill all the time, but the key insight of passive walking is that energy from one step helps to power the next step. Going back to the cockroaches, biologists showed that they bounced slightly at each step, gaining energy that helped to power the leg forward. This partly explains the complexity of our feet and ankles and shows that their compliance is actually a key aspect of walking well.

One group of researchers in the United States had already produced a pogoing robot, which became not so much an interesting novelty as an example of a better approach to bipedal walking.[4] If a robot has compliant—that, is springy—legs, then its own body weight will produce a bounce at some natural frequency. The new generation of bipeds walk much faster and have better stability than those that took the earlier approach of working out on each cycle exactly where to put the foot down.

This approach also reignited interest in applying hydraulics to a robot. Using fluid makes it easy to produce compliance compared to motors and electricity. However, hydraulics have downsides, as well: they need pumps, and if the robot falls over, it risks leaking hydraulic fluid, almost like green blood.[5]

Some impressive videos show the capabilities of the current generation of walking bipeds.[6] Remember, however, that videos represent a best case and do not tell us much about reliability or failures; and in many—though not all—cases, they show robots that are being teleoperated rather than acting autonomously. In the United States, the Defense Advanced Research Projects Agency (DARPA) runs a regular challenge competition focusing on the abilities needed for search and rescue operations, and this is a better test of the current state of the art in biped robots.

Postcompetition performance assessments by some of the participating teams pull out factors that might not occur to you.[7] Some robots performed badly because the challenge environment was just slightly different from the one in which they had been tested. For example, the bit on the power tool the robot had to use was slightly longer, or the walking surface had a different friction from the home lab. Robot motors would overheat and stop working properly.

The set tasks had to be completed within an hour, putting human operators under a lot of time pressure. Operator errors were one of the main causes of robots falling. Many fell or got stuck, whether because of operator errors, errors in the robot, or hardware failures, and had to be rescued by human team members. It turns out that roboticists have conducted much more research on bipedal walking than on how a biped can get up again if it falls over. In summary, we have made a great deal of progress, but bipeds are not yet ready for real-world tasks, especially critical ones.

This is why you will see a lot of indoor robots with wheels, and outdoor robots with tracks like a small tank, or multiple legs. Wheels hugely simplify the task of robot movement, though they also restrict its scope. A wheeled robot has only two degrees of freedom: it can move in the direction it is facing (translation) or rotate to face in a different direction. To make rotation easy, it helps if the robot has a circular base, one drive wheel and two others that can be turned in any direction. In this way, both degrees of freedom can be directly controlled, making the robot a *holonomic* system. This is not usually the case in an automobile, whose steered wheels turn only so far, so that it cannot turn right around on the spot.

Wheels also use energy more efficiently than legs, at least on smooth terrain. We see this when we compare the speed of a human on a pedal cycle with one on foot. Energy efficiency is a fundamental issue, since it limits how far a robot can travel, and in general how long it can function. Whatever its method of mobility, a robot needs a power source. Oddly, few discussions of robots and their capabilities mention this point, but battery technology is actually one of the biggest constraints on robot functioning.

A mobile robot could run off mains electricity, but this involves trailing a cable—a tether—behind it, then not getting tangled up in it. For some applications, a tether is useful; think of a robot venturing into an environment too hostile for humans and then getting stuck there. A tether can be used to pull it out. In general, however, it is more useful for the robot to have its own onboard power supply to support autonomy.

Living things use chemical power sources. Human muscles are driven by chemical products derived from food. Some chemicals are present in the muscles in small amounts that can be processed for immediate effort without using oxygen. Over the longer term, oxygen is used to break down glucose into the chemicals that muscles need. This allows us to store our fuel, in the form of fat. Robots, on the other hand, are nearly always driven by electrical power from batteries.

Batteries deliver electricity quickly and efficiently, but chemical fuels have a much greater energy density. They release far more energy for the same weight and volume. The petroleum used in automobiles has an energy density about five hundred times that of a lead-acid car battery, and glucose has an energy density at least ten times that of a lithium-ion battery, depending on its charge state.[8] Thus batteries in a robot carry much less energy for their weight than we do in our body fat.

Legged robots also use energy much less efficiently than humans. Back in the 1950s, scientists formulated a measure called *specific resistance* (SR) that allowed them to compare the costs of ships, land vehicles, aircraft, and animals traveling at different speeds.[9] SR is a ratio of the power being used to weight and velocity. Later work shows that a horse trots with an SR of 0.2, and a human walks with the same SR, but the best biped robots, using efficient passive walking approaches, still only attain an SR of 0.7 or so, more than three times as much.[10]

Furthermore, the motors driving robot movement are far from the only users of energy. A robot's sensors, controllers, communication equipment, and computers also draw a great deal of power. Estimates suggest that more than half the power used in mobile robots goes to these components.

The outcome is that robots use battery power quickly and must be recharged often.

The pioneering biped robots that the Japanese company Honda produced in the mid-1990s, beautiful pieces of engineering though they were, lasted for only twenty minutes between recharges. Battery technology has improved substantially since then, and industrial work on electric cars and mobile phones is pushing this forward. However, working time between charges is still measured in a small number of hours for most robots: from one and a half to four, even for robots with wheels rather than the more energy-demanding legs. Though recharge times have dropped, they are also still measured in hours, and during recharge the robot cannot move. Were these constraints mentioned more in public discussions of robots, the idea that robots might somehow soon take over the world might seem less plausible.

We now see a great deal of research in how to make robots more power efficient. Researchers are also trying to develop new power sources with greater energy densities. An example is the *sugar battery*, based on the way sugar releases energy in cells.[11] One of the obstacles being tackled is that turning sugar into energy in living organisms involves a complex process. It uses a train of regulating chemicals called enzymes. If a sugar battery is to be reused, the enzymes must remain in it when more sugar is added. It also needs an air source, since the process requires oxygen; the battery cannot be sealed in the way lithium-ion batteries are.

Greater energy density is not the only advantage of a sugar battery. Sugar is more widely available than lithium and much less poisonous. It does not have the same recycling problem. Its power output is consistent, whereas lithium-ion batteries drop their voltage as they discharge, potentially producing anomalous behavior in a robot using them. A sugar battery's outputs are water and electricity. If such power sources become commercially available, they could make a big difference to the usability of robots in real-world applications.

We saw that the design of legged robots benefited from the study of how living creatures walk, even if the resulting robots are not identical to the real thing. The application of biology to engineering is known as *biomimetics* where ideas are copied, and *bio-inspired* where general principles are used. Both approaches have been applied to other forms of robot movement: flying, swimming, jumping, climbing, and crawling. These are not robots intended to live with us in the sense of sharing our social spaces. Instead they take other creatures as their inspiration and try to take on some of the abilities of birds, fish, lizards, insects, or snakes.

Inspired by aerospace engineering rather than biology, both an airliner in fly-by-wire mode and a cruise missile can be thought of as flying robots. But what has made flying robots really popular is the advent of cheap quadcopters, in which four propellers are each attached directly to a motor. As hobbyist vehicles affordable by many enthusiasts, quadcopters are normally flown by teleoperation.

An autonomous flying robot needs onboard processing on top of the batteries, camera, and communication hardware it already carries. A bigger payload in turn needs better motors, using more power. As it is, quadcopter flight time is measured in minutes rather than hours: somewhere in the range of six to thirty minutes currently. Putting in bigger batteries makes the quadcopter heavier and therefore even more power hungry. A small internal combustion engine is more efficient, but also more expensive, and produces a bigger quadcopter. Without its rotors turning, a quadcopter drops from the sky as a helicopter can, so it is inherently energy demanding.

A fixed-wing flier is more economical, since it can conserve energy by gliding, but the lift that keeps it up is related to wing area, in square meters, as well as to a pressure difference derived from its velocity. This means that lift drops exponentially as wing size drops: it is why large jumbo jets take off so much more slowly than small aircraft and why aircraft designed as gliders have much longer wings. Size was one of the reasons why birdmen like the one at the start of the chapter failed: the wing size they needed to lift their mass was much larger than they realized, and could never have been flexed by a human-size shoulder. Though model-size fixed-wing planes are small and have little mass, they still have to travel relatively fast to maintain the necessary lift. Hobbyists prefer quadcopters because they can hover and take interesting photo footage of the ground, which would be flashing past for a fixed-wing flier.

However, birds have both lower power use and higher maneuverability, as do bats and flying insects. They also adapt to windy conditions that blow small flying robots about. Researchers have studied all these examples as a basis for better robot fliers.[12] A German group produced an autonomous bird robot modeled on the herring gull in 2011, and a number of research groups are working on robots based on hummingbirds, known for their especially agile flight.[13]

Like the human skeleton, the biological structures of flying animals are often more complex than engineers today can replicate. A real bat wing has forty joints, so that a bat can control both wing angles and the rigidity of the membrane of which the wing is made. The design dilemma is how much to take from the real animal: enough to gain some of the same abilities, but not so much that the resulting structure is impossible to control. A bat bot is an example of a *soft robot*: its wings are made of deformable cloth-like material, so that it is potentially much safer around people than the metal and propellers of quadcopters.

We have seen that it is premature to think of biped robots freely striding around in the near future. In the same way, swarms of bees or other flying insect robots remain the stuff of stories and TV shows. Building the mechanisms, controlling and powering them successfully, and dealing with the need for miniaturization all present big engineering challenges. However, scientists and engineers are actively researching birdlike and bat-like robots, and some are beginning to arrive at the stage of start-up companies, even if their products are still very much prototypes.

Just as birds and bats inspire work on flying robots, so fish offer a source of new ideas about swimming robots. Like birds, fish are extremely agile, with much greater maneuverability than conventional underwater vehicles. Most fish can turn very rapidly, and predator fish accelerate extremely fast from rest. They are highly efficient, losing less energy through drag against the water than a propeller-driven vehicle. Research suggests that when a fish's tail creates a vortex in the water by swishing in one direction, it is able to use some of the vortex energy to swish the other way. But again like birds, fish present a major challenge to engineering. They move by undulating, which requires a flexible robot, made of many segments while still remaining waterproof. This is easy to achieve in the tiny clockwork fish sold for children's bath-time amusement, but harder with electronics. The usual design is to put most of the electronics in a sealed head and to attach swimming segments, each driven by its own motor.

The first robot fish was the RoboTuna, built in the 1990s at the Massachusetts Institute of Technology. Since then, robot fish have been constructed by groups in the United Kingdom, France, China, and Japan, among others. The ultimate aim is to produce small, efficient, and stealthy inspection fish, but so far the biggest success has been in entertainment, with robot fish tank installations.[14] These are reminiscent of the ancient singing bird automata of Byzantium.

Land-based multisegment robots are much more developed. These use snakes as their biological model. The idea here is robots that can move through small spaces, whether under floors or in pipes, in the rubble of destroyed buildings, or even in the human body as an automation of endoscopic surgery. Their shape provides one advantage, and with multiple segments, they can still function if some segments are damaged. Like trains—or earthworms—they can have sensors and control mechanisms at both ends so that they can move easily forward or backward.[15]

Real snakes do not have wheels, but many snake bot designs do. Sometimes these are passive wheels, whose different friction with the ground in different directions helps to generate the snakelike undulations. Some designs have a motor in each segment to drive its wheels. In some designs, the wheels are replaced by tracks.

The development of snake bots for use in surgery—without wheels—is also an active research area, though not yet used on real patients. Endoscopy already involves running a cable inside a patient, with a camera and tiny surgical tools attached. Giving it some independent locomotion capabilities makes it a snake bot, though it would have to be largely teleoperated by a surgeon.

Finally, bio-inspired robots have also been based on gecko lizards, which can climb walls with their sticky feet, and grasshoppers, which can jump many times their own height. Biomimetic and bio-inspired robots are still nearly all prototypes rather than products, but they show a great deal of potential.

4

SENSES: WILL THEY BE AWARE OF US?

Back in 2002, a small robot called Gaak escaped from the Magna Science Adventure Centre in Rotherham, England. Gaak had been part of a "living robots" exhibition and was taking part in a "survival of the fittest" competition. Gaak trundled out of its pen through a small gap. Then it went down an access ramp, exited through the front door of the center, and was found at the bottom of the drive, where a visitor nearly ran it over in his car.

That, at least, was how the story was reported in the press.[1] The real story was entirely down to its sensors. Gaak had been programmed with a behavior known as *taxis*—which has nothing to do with automobiles and is pronounced *tax-iss*. The idea is that you turn toward the strongest signal on your set of sensors and move toward it. Gaak was doing *phototaxis*, moving toward lights, because it was supposed to act as a "predator," seeking out smaller but more nimble "prey" robots, each with a light source attached. When it caught up with one, it would plug itself in and "eat" the prey's electric power. But the day was bright and sunny, and Gaak had been left switched on, so it headed toward the sunlight beaming into the center.

As it happened, there was a large tree at the bottom of the drive, with its branches moving gently in the wind, producing a moving pattern of light and shade beneath it. Gaak got stuck there as it tried to follow different patches of light and went round and round in circles.

Two interesting points emerge from this story. The first is that robots need sensors to react to a world in which things change. The second is that reacting sensibly—"doing the right thing"—is not always straightforward, and linking robot actions to sensor input can produce unintended results, especially if a robot moves outside the environment for which it was designed.

The word *robot* is still used for mechanisms that have no sensors, as with many generations of industrial robots. However, we argued in chapter 2 that for something to be rightly called a robot, it has to be able to respond to changes in its environment in real time. This means it has to have sensors. As we will see in the next chapter, the first thing a robot wants from its sensors is information that allows it to avoid obstacles rather than crashing into them as it moves around. Working out where it is and how to get where it wants to go—localization and navigation—is the next thing it needs to be able to do.

Just as we typically imagine robots in humanoid form, we also imagine their sensing to be like our own. After all, a robot with a camera is seeing what we see, isn't it?

But of course it isn't. Just as we found with human versus robot bodies, we cannot equip robots with senses just like our own. The human visual system is far more complex than a mere camera and, like our bodies, not completely understood. A camera has standard light receptors spread evenly across it. The retina has two types of receptors. Rods work at low light levels, do not process color, have lower visual acuity, and are more sensitive to motion. They are distributed toward the edges of the retina. Cones are concentrated more toward the middle of the retina and process color with higher visual acuity. The resolution of the retina varies, with its highest resolution in one central area only populated by cones, called the *fovea*.

While a camera points at a whole scene and software processes what it captures, the eyes are always moving. Our light receptors saturate if the same light falls on them all the time, so that if our eyes did not constantly move, we would not see anything at all. Very rapid eye movements called saccades allow the fovea to be moved to the "interesting" bits of a scene. The brain edits fovea data out during this movement. It also steadies the visual field when we are moving the rest of our body. Moreover, the retina captures images upside down, and the brain turns them right way up. If you were to wear glasses that turn what you see upside down, the brain would eventually adapt and invert it back—until you took the glasses off, when the brain would have make this adaptation again.[2]

When we look at a scene, we see objects. When a robot points a camera at a scene, it receives a set of numbers, one for each *pixel*, or picture element. This is more like standing very close to a large billboard and seeing each printed point on it—except that we see a printed point with a color, not a set of numbers representing it. The robot is getting raw numerical data, when what it needs is information about what is around it.

The number of numbers the robot receives depends on the resolution of the camera: for example, it might be providing four megapixels, or over four million numbers per image. How does the robot extract the objects we might see in the same camera image? The first thing is to look for *features*, ways of grouping pixels together. A standard approach is to look for pixels with similar colors and intensities and to form them into *edges*, or lines. The lighting in the scene is bound to have a big effect on how this happens. Imagine a person standing in front of the robot, side-lit by sunshine coming through a window. The pixels down the illuminated edge of the person's face will be a different color from the edge in shadow, and somewhere in the middle of the face, a shadow line, which has little to do with the face itself, will divide the lit from the unlit side.

One thing that helps in processing an image is having an idea of how far away the pixels are. If the person standing in front of the robot has more people behind them, we do not want pixels of similar colors belonging to more distant people to be added into the edges found for the near person. Our eyes are binocular, giving two images for every scene, and the difference between these images, owing to the separation of the eyes, gives a displacement that the brain can use to judge distance. The eyes converge more for a close object than they do for a distant one, and the eye lenses have to change shape to focus on objects, so that the muscular tension also gives us an idea of how far away objects are.

We could produce some of these effects by giving the robot two cameras. An easier approach is to use a camera with a range-finding capacity. Rather than just letting light fall on the camera, we can scan the scene with a low-power laser and collect the bounces. Those from near objects come back much more quickly than those from distant objects, so that we can add a distance measure to each pixel. This is not at all like a human eye, but it gives the robot extra information about which pixels to form into an edge belonging to one object, not several.

Lasers are relatively expensive but have good accuracy unless they hit a highly reflective surface, like glass. They cannot be used for autonomous underwater vehicles (AUVs), because the water scatters the beam, making it far too short-range. Cameras are problematic as well for AUVs because the deeper you go, the less light there is. AUVs use ultrasound instead for range finding, similar to the sensing of dolphins. The downside is that ultrasound spreads out much more than a laser as it travels, so that when it bounces back, it usually includes lots of ghost echoes.

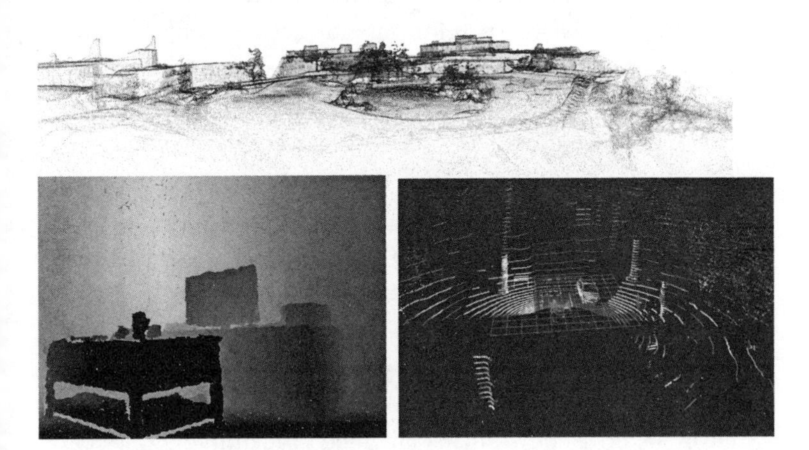

4.1 Three graphical renderings of sensor data: (a) *top*, Stirling University from a drone, a point cloud, where each point is a number that must be processed; (b) *bottom left*, front lounge using a Microsoft Kinect sensor processed to show planes; (c) *bottom right*, road, car, trees in laser slices from lidar sensor. You can make out objects in these perhaps; that is your eyes doing clever stuff.

This takes us to the hidden assumption in the story so far: that the numbers coming from sensors are accurate. In reality, the data from sensors include an element of what engineers call *noise*. This was originally the term used to refer to the hiss—static—you can hear on an AM radio signal and was then extended to refer to any unwanted electrical fluctuations. Noise will produce small random changes in range data, even if the robot is not moving at all. Image noise produces similar changes in the brightness and color of a camera image. These variations come from many sources: some internal, caused by the movement of electrons within the sensor itself, and some external, from the environment. Nearby electrical devices generate magnetic fields that can induce currents, and even solar radiation can produce electronic noise.

Sensor data therefore contain an unavoidable element of uncertainty, making such information accurate only within certain limits. So a moving robot must distinguish between changes in its range data caused by distances actually changing and those caused by noise. Engineers have produced what they call *filters* to reduce sensor noise and smooth out fluctuations, so that the data the robot is using to assess its surroundings are reasonably accurate. We'll look at this again in the next chapter.

Let's get back to the edges, working on the assumption that noise has been successfully dealt with. Edges are only the start of the story. The robot needs to work out what object the edges are edges of. Humans are good at identifying objects because we have a great deal of knowledge about the world around us and the objects it contains. Babies are predisposed to seek out face-like objects, suggesting some built-in knowledge. Any of us can identify an animation of moving lights on the joints of an invisible graphical skeleton as a human walking and even specify the gender. Babies spend a great deal of time looking at the world and learn about objectness quite quickly—a lot of it in the process of developing the hand-eye coordination needed to grab things or maybe bash them with a spoon. Giving a robot a set of edges from an image and then asking it to identify the objects, if they could be anything, is not the same sort of problem.

Making a set of edges into an object is less challenging if the software already has expectations about what it might be, or is looking for something that fits a particular pattern or model. Much modern visual-processing technology has been developed in the context of surveillance, where a camera should focus on people or their faces. Software can hold a sticklike model of the human body and then try to fit what it sees to that model. If this works, then the model can help predict what the camera will see next. It also helps the camera to follow a specific person if that person passes behind others or is otherwise partially hidden. In the same way, a system can hold the pattern for a specific group of features. A face includes an oval edge, eyes two-thirds of the way up, and a mouth near the bottom.

Face recognition software works well using these features as long as the lighting is not too different from what the software expects. It will try to match specific features from the image to the same features in a database of mug shots. Eye shape is particularly distinctive, which is why automatic passport gates work better if people remove their glasses.

We can identify some important differences between what a robot needs to do and what a surveillance system does. A robot is moving around in the world, so it has to be interested in more than faces, especially if it has to interact with particular objects. Maybe it needs to find a mug of tea and bring it to a

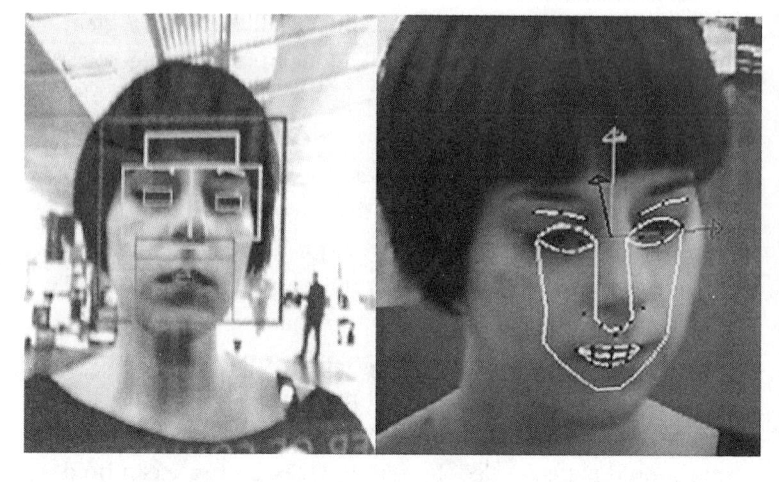

4.2 Software can extract facial features if it has a model of faces telling it where to look.

user or locate a mislaid pair of spectacles. Or it is a fruit-picking robot and needs to find the ripe apples on a tree. For each object, the robot needs a database model or a feature pattern it can use for recognition.

Fortunately, a robot may not need to know what absolutely every object around it is. Obstacles are obstacles, whether table, wall, or desk. The important thing is to avoid them, whatever they are. This is one good reason for giving a robot extremely specific information about which objects it should recognize: any given environment is likely to contain only a few. This also minimizes the computational effort involved in doing the recognition.

The four-million-plus numbers the robot needed to process was just for a single camera image. In a world that is changing, if only because of the robot's own movement, one image is not all that useful. So its camera might be providing at least twenty images a second, possibly as many as fifty. These images constitute a lot of data and require a good processor and substantial amounts of data storage. Holding all of this in the robot takes us straight back to the power issues of the last chapter, not to mention the heat generated and the extra weight. Even robots with a high degree of autonomy may have to pass camera data to off-board processing facilities. Tracking moving objects—like the ball in robot soccer—is usually closer to five frames per second than twenty by the time the processing is done.

As the robot moves, the lighting conditions change, making it harder to take account of their impact on feature extraction. Some of the features may belong to independently moving objects, like people walking about. But the camera is moving too, with the robot, and as we will see in the next chapter, establishing where exactly the robot is after a series of movements can be tricky—not to mention any wobble in the camera mounting and wheels, or worse still, the complex movements generated by legs. Extracting features when the image is bouncing around is unlikely to work well.

The news isn't all bad. What helps is the idea that the scene in which the robot finds itself is not changing much. Features present in one image are likely to be in subsequent images, even if they have moved and changed a bit. The robot also has more control over the changes in the image than a typical surveillance camera that can only zoom or pan: it can move toward interesting features or follow features that are also moving.

The other good news is that because surveillance has been such a big research area, publicly available libraries are carrying out the necessary processing. This allows anyone equipping a robot with a camera to use robust software without having to write it all from scratch.[3] Such libraries are now adding the newest methods in visual recognition, using machine learning, discussed in chapter 8. The idea here is that presenting a learning program with large numbers of images containing the desired object will allow it to learn the best features for recognizing it. If the same object is now presented in a new incoming image, the system ought to be able to recognize that too, without all the work needed to extract the features. Anyone who has dealt with a log-in security feature asking which images in a grid have cars in them is helping to train such systems. Today, though, the approach requires more computational resources than most robots have on board and works in real time only for specific objects—like finding the ball for a soccer-playing robot.

In the end, the current story with cameras on robots is much like the story of mobility in the previous chapter. In a known environment with a limited number of known objects, recognition can work well enough for particular tasks much of the time, depending on the lighting. If the robot has to recognize a small number of people, it can be equipped with their mug shots and can move until it gets a good view of their face. In general, though, a robot sensing with a camera is still rather unreliable and not very fast. It is not in the same league as human visual perception.

However, there is no need to equip robots only with the type of sensors humans have. Just as we can add range finding to a robot's camera, so we could equip it with an infrared camera. This would give it night vision without the need for goggles. Robots nearly always have much simpler sensors than a camera, as well. A bumper with a contact sensor tells the robot if it has actually hit something. Sensors that just do range finding—without trying to build up an image—give the robot information *before* it hits something. These could use lasers, as in the camera we talked about, but bouncing infrared or ultrasound off the surrounding environment is cheaper and may do the job well enough. Robots usually have these sensors spread around the lower part of their body to allow them to avoid collisions. The advantage of simple sensors is that they generate much less raw data and can be processed much more quickly. Sensors of this kind also gain less information about the world, but that means they are less ambiguous.

Robots with wheels often have a sensor called an optical-flow detector. This measures how fast pixels are moving across it, again without trying to build up an image. If you know how far off a level floor an optical-flow sensor is located, this flow gives a good indication of how fast and in which direction the robot is moving.

The simplest sensors are like the one the escaping Gaak was using. These just measure the level of some signal, for example, light, infrared, or sound. You might think this is far too simple to produce any interesting robot behavior. But taxis, whether positive (toward a source of the signal) or negative (away from it), can in fact produce complex interactions between multiple robots, their environment, and one another, as we will see in chapter 9. Many researchers have modeled small robot ecosystems like this, in some cases including a charging station with a light to which the robot is attracted when an internal sensor tells the robot that its power is low.[4]

What about equipping a robot with hearing? We are becoming used to having disembodied voice-driven systems in the home, such as Alexa, Siri, or Google Home, and have come to expect a robot we can call over and talk to. We will look at language interaction much later, but if it is going to work at all, the robot must have facilities to pick up sound. However, just as a camera does not work like human sight, a microphone, or an array of microphones, does not work like human hearing.

Sound is unlike light in being much less directional. We see what falls on our retinas, and move our heads to see other things. Sound can be generated anywhere around us, and we will still hear it. As with eyes, having two ears gives some indication of location, since sound arrives at one ear slightly before it arrives at the other, if the sound originates to one side of us. Sound frequency also provides clues about a moving sound source; think of the change in the siren pitch of an emergency vehicle as it comes toward us and then moves away. Finally, the intricate shape of our external ear is not purely ornamental but acts as a baffle to incoming sounds, helping to localize them.

An array of microphones gives a robot some capacity to locate the source of a sound. But there are still two big problems. The first is that the robot itself generates sound, especially if its motors are operating. This reduces the chances of picking up external sounds accurately and is why human-robot interaction experiments often involve the human wearing a microphone. While people will do this for an experiment, it is unlikely to go down well in more natural environments.

The second problem is more fundamental. Humans, like many other animals, are equipped with an *attentional* system. This means that we do not focus on everything arriving via our senses but only pay attention to the signals that interest us. The structure of the eye, with its high-resolution fovea, supports this process. In the case of sight, law officials collecting witness statements have cause to rue what witnesses see but do not register. Ingenious experiments demonstrate lack of attention—like the one where a receptionist ducks behind the reception station and a different person pops up. Many people fail to notice this.

With hearing, attentional systems allow us to filter out what we call *background noise*: everything we do not want to listen to. Many of our normal social environments contain a great deal of background noise. Doors slam, cars drive along the street, the washing machine rumbles, the radio plays. And yet we can talk to somebody else and, as long as the background noise is not too loud, just cut it all out and focus on what the person is saying. Even in an extremely noisy environment, we are likely to hear—and attend—if somebody says our name. This attentional ability is so important that it has a name: *the cocktail party effect.* At a party, we talk to the group we are with and filter out the sounds of all the other people talking.

As with many of our sensory capabilities, we do not understand exactly how we do this, and it has so far turned out to be extremely hard to implement in a robot. In an everyday human environment, a robot is not good at focusing on a voice rather than all the other sounds hitting its microphone. Indeed, by calling it "a voice," we are already making a strong assumption that it is a single sound stream that we can focus on. But this is what the robot really needs to do if it is to apply speech recognition to it. Once again, there is a big difference between a robot moving around an everyday environment containing several people—for example, an office—and a single person sitting in front of a computer and speaking through a microphone. A disembodied conversational system remains in one location in a room.

It would also be useful if the robot could detect and respond to significant nonspeech sounds, like a doorbell, a wake-up alarm, or a mobile phone ringing when you have left it lying somewhere. These are tricky problems to solve, and researchers sometimes deal with them by linking the robot to a smart building that can transmit messages about events the robot should notice. So when the doorbell is pressed, an attached sensor is activated, and the smart building sends a "someone at the door" message to the robot. A refinement would be to put a camera at the front door and perform facial recognition at that stage, then tell the robot who the system thinks is there.

In this approach, the robot is no longer a single autonomous entity, working in a passive environment in the way a human would, but one element in what is now called the Internet of Things. The robot, the front door camera, the cooking and washing appliances, the lights, and other home items form a single network in which data can be transmitted to whatever part of the system needs it. Before anyone leaps to the conclusion that the robot can now go anywhere and do anything, it is worth pointing out that the smart environment has the same sensing limitations as the robot and is far from able to interpret the data it receives as a human does. It also means that the robot is bound tightly into the smart environment, and its ability to operate autonomously without all the extra information may be limited.

All of this underlines a point we made at the start of the chapter: a robot's sensing is usually tailored to the environment in which it operates and may work badly, unpredictably, or not at all when the robot enters a different environment.

We have discussed two of the five external human senses: sight and hearing. What about the other three: touch, smell, and taste? We will discuss touch in chapter 6, so how about smell and taste? In humans these senses are linked in that taste itself, handled by taste receptors on the tongue, deals only with seven basic categories: bitter, salty, sour, astringent, sweet, pungent (e.g., chili), and umami, the last having been added to the set quite recently. Much of what we think of as taste is generated by our sense of smell.

Robots do not eat, so perhaps they do not need taste? On the other hand, smell contains useful information: tracking environmental spills or noticing a gas leak or bad-smelling garbage that should be emptied, for instance. A robot can smell only if it has an appropriate sensor, familiarly known as an "electronic nose."

Just as a camera mimics some aspects of the eye, and a microphone some aspects of the ear, so an electronic nose is intended to parallel some of the functions of the nose in animals, including humans. The human sense of smell plays much less of a role in our sensory life than it does in, say, a dog, but it is still more capable than at first supposed. Some research suggests that the human nose can detect around a trillion odors.[5]

A specific skin layer high inside the nose is loaded with smell receptors, about 450 different types. The molecules forming a smell—a combination of chemicals—bind to specific receptors, more or less strongly. Each binding sends an electric pulse to a part of the brain called the olfactory bulb. This distributes the signals to other parts of the brain, so that the odor is recognized—and again, although we know which parts of the brain are involved, we do not entirely understand how this is done.

So an electronic nose needs a way of distinguishing the chemicals in an air current. We might use a variety of gas detectors, whether incorporating gas chromatography, spectroscopy, or organic compounds that react in a particular way to the gases involved. The range of gases that can be detected in this way is much lower than the human nose manages, but different combinations can be profiled to identify different smells. A robot can use a pair of such sensors for taxis, thus allowing it to find the source of a smell or perhaps a gas leak. Like seeing and hearing, smelling becomes much more practical and useful when it is designed for specific environments with specific smells.

A much simpler system was used for a gasbot built by Swedish researchers that looks for escapes of methane in landfill sites.[6] This robot projects a laser thirty meters ahead of itself at a frequency that is absorbed by methane. The bounce back is analyzed to see how much was absorbed, giving a measure of the methane concentration.

The fundamental problem of robot sensing is clear from all the modalities we have looked at in this chapter. Sensors provide numbers; what the robot needs is *information*. The information it needs depends on what it is trying to do: avoid obstacles, navigate somewhere, recognize objects, track people, engage in language interaction, react to specific sounds and smells in an appropriate manner. Our own senses provide us with information in such a seamless way that it is hard to understand just how much work is needed to produce sensory information in a robot and how much more limited than human sensing the outcomes still are.

5

THE LOST ROBOT: COULD THEY KNOW WHERE THEY ARE AND HOW TO GET HOME?

Where are you right now?

Generally, we can always answer this question. We worry if we don't know where we are and say we are *lost*. But there is more than one way for you to give me an answer.

I'll answer the question myself. I am in Edinburgh, Scotland, according to Alexa, when I just asked. But I could be in Edinburgh and still feel lost if it is a part of the city I do not know and I am not sure which way will take me to somewhere I do know.

Google Maps reckons that I am at an address two doors down from mine. Close, but no cigar. It also told me what my latitude and longitude were, though I had to ask via my phone, which I allow to record location. When I asked via my laptop, it had no idea.

My own feeling about where I am is that I am sitting at my kitchen table. Later I will be in my bedroom, hopefully asleep. But if someone in the United States asks me on social media where I am, I might tell them "in Scotland," or if I thought the person might not know where that was, "In the UK." If I replied with my latitude and longitude, that person might be taken aback.

So here we are, back at the distinction between information and data we finished with in the last chapter. This is not coincidental, since my location has been derived from sensor data. The information people need is related to how they will use it, which is why there is more than one way of explaining where I am.

How *I* will use this information, a little later on, is to get up from my chair, avoid crashing into the chair or the kitchen table, navigate myself through the kitchen door into the hall, down the hall without hitting a bookcase, a linen chest, or a filing cabinet, and through the correct door into my bedroom. Knowing I am in Edinburgh, in Scotland, on planet Earth, in the solar system (and so on), or indeed my latitude and longitude, does not help with these tasks. What I need for local navigation is a local map, the ability to use it, and the coordination to avoid obstacles as I do. A robot needs similar competencies and without them is unlikely to be at all useful.

We saw in the last chapter that robot sensing is greatly inferior to human sensing. This is one reason why robots find localization—establishing where they are—and navigation much harder than a human would. Earlier we touched on another reason: a robot gets numerical data from its sensors, but it needs to convert those data into useful information for these tasks. Robots in films almost never struggle with such skills, but their deficits are one reason why filmmakers have never used real robots. People who fear robots taking over the world may find that their assumptions about how a robot can localize and navigate successfully are way ahead of what a real-life robot can actually do.

An early approach was to give the robot a map of its location, usually a floor map, as if drawn on graph paper with *x* and *y* axes. An arbitrary 0,0 point was allocated, say, in one corner of the map. To localize, the robot needed to know its *x,y* position. Once it knew this, then it could navigate to a new *x,y* position by working out the amount of straight-line movement needed (translation) and the number of turns needed (rotation). Well-known search algorithms could be used to decide an optimum path in terms of translations and rotations, and the robot could then follow it.

This sounds straightforward. The approach does work well for nonplayer characters in computer games, but in the real world, things are not so simple. First, the map may be inaccurate. While walls do not move around, chairs and desks may do so, not to mention the movement of other occupants within the space. If something stands in the way of its planned route, what should the robot do? What if the robot is in an environment where a map is not available—for example, a search and rescue robot in a collapsed building?

Second, we saw in the last chapter that the robot's sensor readings contain an unavoidable element of uncertainty. Finally, the robot may not follow the desired path accurately, either.

For a robot with wheels, the number of times the wheels have turned (known as *odometry*) should provide its translation. Adding in the record of the rotations it has made, the robot should be able to plot its position on the map relative to where it started. This approach is called *dead reckoning,* since it relies purely on the robot's internal data. It is yet another reason why so many robots have wheels rather than legs.

Unfortunately, real wheels slip a little, and actual rotations may not be exactly the ones commanded. Errors may be small, but they accumulate, so that the longer the route, the less likely the robot is to be where its calculations say. Inside a building, this could mean the difference between successfully following a path through a doorway and getting wedged against a wall that the robot thinks should be the doorway.

The robot therefore needs sensor data about the environment so that it can check its internal calculations. The floor does not have *x,y* coordinates written on it, so the robot needs to find some feature on its map, identify and locate that feature in the real world, and compare the two. Let's say the robot picks the bottom right corner of the doorway and uses its range finder to determine how far away and at what angle the corner is. Comparing this actual translation and rotation with the ones its calculation says it should see will allow the robot to correct its position, a process known as *recalibration*. A complication is, as we saw in the last chapter, that these sensor readings also contain some uncertainty.

Was there a way, researchers wondered, of working out how much error had accumulated, so that the robot could recalibrate at sensible intervals? Knowing the exact location error is equivalent to knowing the correct position, which is the problem to be solved. However, it turns out that we have methods that allow us to statistically estimate the degree of uncertainty about the position. This is not quite the same thing, but it effectively draws a boundary around the robot and says, "I am somewhere inside this area." If the area is not too big, then the uncertainty might not matter that much. After all, we do not know our exact *x,y* position, but we still manage to navigate successfully.

We mentioned filters in the last chapter as a way of dealing with the uncertainty of sensor readings. The one most widely used for estimating the uncertainty in the position of moving objects—not just robots but also spaceships and cruise missiles—is the Kalman filter.[1] The Kalman filter combines two uncertain pieces of information, the information from external sensors and the internal model of what is happening—the velocity—to get a better estimate than either provides on its own.

Each of these can be represented as a statistical distribution, with a mean giving the height of the distribution and a variance giving its width. There is a greater likelihood of the robot being close to the mean than somewhere in the tail, and by combining the two distributions, we fuse the two sorts of data into a better estimate of where the robot is and what its velocity is. The current estimate is called the *state*, and what the filter does is to use this to estimate the next state. In doing so, it can add in any changes the robot makes to its speed and rotation, as well as an estimate of the impact of wheel slippage and rotation errors. New information from features in the environment can shrink the combined distributions, reducing the degree of uncertainty. So though the robot does not know exactly where it is, it does know approximately where it is. Since the filter uses only the most recent observation along with its most recent state estimate, it does not have to store a lot of data and can run in real time.

In the 1980s, some researchers moved away from the idea that a robot needed an accurate map and a complete plan of the route. They observed that many animals navigated successfully over long distances without either—think of migrating birds, for example. Rather than having detailed global knowledge, animals could extract information from the world around them that guided their behavior successfully. The slogan of this school of thought was "the world is its own best model,"[2] and it is often known as *behavioral robotics*.[3]

So if the robot knows in which direction its goal is located, it could move toward it without a complete plan and need not know exactly where it is located as an *x,y* position. The robot could head for any features in its environment that lie in the right direction. This method is much closer to the human take on navigation, which tends to be based on landmarks. Think of asking a stranger for directions; you would expect something like "head through that door, turn left down the corridor, through the swing doors, and the second door on the right."

This approach is more flexible than working out a complete plan, but it can lead a robot into trouble. Say the robot is in an office, and the location it is trying to reach is indeed way down the left end of a corridor. Say the office has a door into the corridor. Picking the left-hand corner of the office as the best direction to go will mean the robot gets wedged. It has to exit through the door before it makes sense to turn left, as in the human directions. So some combination of planning, providing useful waypoints like doors, together with some behavioral improvisation in between waypoints, may work better. This is how the map-based software that gives directions to car drivers works: it focuses on the turns the car needs to make while leaving the driver to handle the short-term navigation. We will come back to how this might work for autonomous cars later in the chapter.

It occurred to researchers that locating useful features in the environment for navigation meant that the robot was actually making a map as it went along. This map was a robot's-eye view, as it were, but since it was built from current sensor data, it was likely to be more accurate than one given in advance. Maybe mapping and self-localization need not be independent tasks but could be seen as two aspects of a single problem. This insight led to the most successful current approach to navigation, known as SLAM—simultaneous localization and mapping.[4] Conceived in the 1980s and 1990s, it has been widely applied to robots from commercial household vacuum cleaners to autonomous cars.[5]

SLAM really came to prominence when the autonomous vehicle Stanley used it and won the DARPA Grand Challenge in 2005. DARPA had set up the Grand Challenge in 2004 with a 150-mile course across the Mojave Desert.[6]

In the competition's first year, the best entrant managed only just over seven miles of the course before getting stuck, prompting intensive effort by teams for the 2005 rerun. As a result, five vehicles finished a difficult course that included switchback roads, a mountain pass, and three narrow tunnels. Stanley, the entrant from Stanford University, completed the course in just under seven hours, ten minutes ahead of its rivals from Carnegie Mellon University. This was an impressive achievement after the previous year, though you can see that none of the vehicles were traveling very fast. The outcome was responsible for making SLAM the dominant approach to navigation.

SLAM can be implemented with a variety of algorithms. The robot needs a range-finding sensor such as a laser, binocular cameras, or sonar. It cycles through the five steps of extraction of landmarks, data association, state estimation, state update, and landmark update.

Using the word *landmark* may make you think of the kind of thing a human would use: swing doors, a gas station, a stairway, a park. We take for granted the ability of our fellow humans to extract meaningful objects from what they see. While robot landmarks need to be static (no moving objects) and unique (not to be confused with other landmarks), they are far more likely to be pieces of geometry, like corners, than something a human would pick. For example, a spike in laser range data, where neighboring values are a substantial distance farther away or closer, might be a landmark. There is no point in adding the complexities of object recognition into navigation.

The basic idea is to extract a landmark, move, and then associate the new range data with the same landmark from the new position. This gives a position for the robot, which can be compared with its position from odometry and rotations. The robot uses this information to make a new estimate for its position, which could be carried out by a Kalman filter (or another estimator). The state is updated, and then new landmarks that have been detected are added to the set being used.

Things can still go wrong. There might be no landmarks at all in a specific time step. A landmark could be extracted once and then never seen again. Worst of all, when previous landmarks are sought after a move, the wrong one might be matched. If this happens, then the robot's estimated position will be completely wrong. Lots of work has gone into finding ways of dealing with such problems so that SLAM is robust enough for commercial use.

We have been speaking of a map as if it were like the maps we use. The discussion of landmarks shows that this may not be so. More than one way of representing a map exists for a robot, depending on the type of environment and the navigation task at hand. A wheeled robot indoors could have a map that divides the space into squares and allows the robot to decide, by using its range finder, which squares are empty space. Or the robot could have a graph made of nodes for waypoints with distance and direction information connecting them. Or it could have labeled edges or a set of landmarks with distances between them.

Things are different for outdoor robots. For one thing, they can use data from global positioning satellites (GPS). Or at least they can as long as they can see three satellites, allowing them to triangulate their position. This does not remove the need for a filter—back to the Kalman filter again—since GPS sensors also have noise. If you access the raw data for the GPS sensor on your mobile phone, you will see that the values wander around even when the phone is absolutely still. You will also find the signal mostly absent inside buildings, and it can also be blocked outside by tall buildings, trees, canyon-like city streets, and other obstacles. GPS data cannot substitute for the range-finder-based information already discussed, and indeed the Uber prototype autonomous vehicle had some forty-eight lasers on board.

Other outdoor factors make the task for a robot vehicle more difficult than for many indoor applications. Everything moves much faster, giving less time for decisions. Whereas the Mojave Desert had almost no other traffic, cities are full of other vehicles, cycles, and pedestrians. For this reason, an autonomous car must be equipped with a knowledge of traffic laws so that it will stop at traffic lights and give way correctly at intersections.

However, every human driver knows that it is a mistake to assume that other road users are rigidly following traffic laws. Trucks double-park while unloading, taxis make sudden U-turns, cars pull out without checking their mirrors first, drivers throw their door wide open and step out into the road from the car they just parked. Pedestrians—at least in the United Kingdom (not to mention China or Brazil)—decide they can cross the roads at arbitrary points rather than at traffic lights or on marked crossings.

Drivers follow informal rules too, which vary in different countries. In the United Kingdom, depending on context, drivers flashing their headlights may be trying to tell other drivers they need to put theirs on, or may be indicating that they are giving way when, strictly speaking, traffic rules say they do not have to. All of this means that proximity sensing is crucial in these environments, another reason for making sure the autonomous vehicle has plenty of range-finding sensors.

Avoiding obstacles is important for robots inside buildings, but it is much more important for autonomous vehicles. Inside a building, a robot can usually avoid a static obstacle by using its range finders to detour around it. Usually it does not need to know what the obstacle actually is. Detouring may not be feasible for an autonomous vehicle, or it may only be possible by breaking a traffic law and crossing to the wrong side of the road.

For example, if an autonomous vehicle is following a bus in a narrow street, with a lot of traffic coming the other way (a scenario much more likely in countries other than the United States), then it may have to stop every time the bus does and wait for it to move off again. So the vehicle needs to distinguish between taking action to avoid an obstacle and waiting until the obstacle ceases to be one. Frequently the correct action is not avoidance at all but emergency braking, especially if the obstacle is a person. The correct action depends on being able to identify what the obstacle is, which takes us back to object recognition.

Dynamic obstacles—other moving objects—are harder to assess, even in indoor environments. Consider the small electric buggies in airports that carry passengers with mobility problems to and from their gates. These buggies may navigate inside crowded terminal buildings with many people walking toward them or standing in their path. This is why they travel slowly. However, their drivers also expect people to get out of the way and do not try to detour around them. This is a standard tactic when driving through an oncoming crowd of pedestrians. If the driver tried to avoid everybody, the buggy might never reach its destination. Buggies will hoot to alert people apparently unaware of them.

These complications are why driverless vehicles have not yet been deployed outside a few highly specific environments. Airborne drones and driverless train services are the biggest application areas, with the latter varying from shuttles on dedicated tracks between airport terminal buildings to whole metro lines.[7] Driverless tractors are not yet in operation, though they are close to market, and driverless mining machines have been in use for more than ten years in Australian opencast iron ore mines, though they use teleoperation.[8] Common factors here are few or no obstacles, specifically not people, and no or very straightforward navigation decisions. So what does "completely autonomous" actually mean? This is not a purely theoretical question, because regulating the introduction of such vehicles requires internationally accepted definitions.

The navigation task entails a temporal hierarchy in how it is carried out. On the most immediate timescale, it is all about motor decisions: turning wheels and steering, as well as avoiding collisions. On a longer timescale, it is about identifying position and navigating to waypoints. On a still longer timescale, it involves establishing and following the waypoints that will get the robot to a particular destination, and on the longest one, it involves deciding what the destination should be. We can see autonomy as layered rather than an all-or-nothing thing, where the driverless systems just mentioned work in the first two layers.

Functional layering is vital to the design of robots, and we will return to this issue in chapters 7 and 9. However, the regulation of driverless vehicles focuses, as you would expect, on the role—or not—of the driver. The IEC standard used for driverless trains distinguishes four levels, from full driver control upward.[9] In level 2, a driver in the cab is responsible for closing doors, reacting to any obstacles on the track, and handling emergencies, but the train drives automatically between stations. Fly-by-wire commercial aircraft often work at a similar level of autonomy.

In level 3, there is no driver, but a staff member is always on board to handle emergencies. This requires staff members to take over driving if they see the need. Level 4 dispenses with this and can run with no human staff on board. The IEC scale has been refined for driverless cars, starting with a level 0, where the driver controls everything.[10] Level 1 covers cases where the car has autonomous features exercised one at a time, such as cruise control, autonomous braking, and autonomous parking. Level 2 covers the case where two functions are autonomous at a time—for instance, steering and throttle. The Tesla Autopilot sits at this level.

Level 3 is the same as in the IEC definition and is where recent autonomous car trials have taken place, on the basis that the driver can take over at any point (assuming the driver is attending). Levels 4 and 5 split autonomous operation: level 4 covers the ability to drive autonomously in certain scenarios only—convoying on auto-routes, for example—and level 5 the ability to drive autonomously in all scenarios.

There has been much enthusiasm for autonomous cars, but recent trials have shown that real-world environments are more challenging than was foreseen. This has resulted in a number of traffic accidents, the majority so far involving driver fatalities and the Tesla Autopilot. However, a pedestrian fatality in Arizona in 2018 involved a level 3 car and seems to have resulted from an object recognition failure, as well as the failure of the driver to watch the road until too late.

A woman crossed a poorly lit multilane road, pushing a bike laden with shopping bags. The vehicle's object recognition system classified it first as an unknown object, then as a vehicle, and finally as a bicycle, taking four of the six seconds between detection and impact. Each of these required a different action. Then, 1.3 seconds before impact, it determined emergency braking was needed but could not action this automatically, as according to the initial accident report, "emergency braking maneuvers are not enabled while the vehicle is under computer control, to reduce the potential for erratic vehicle behavior."[11] Meanwhile the driver was apparently looking at their phone and braked only after the impact.

The comment about automatic braking is an interesting one, since it suggests a sensitivity problem that might cause the vehicle to emergency brake in situations where a human driver would not. Since emergency braking can also cause knock-on accidents, and certainly alarms the driver, getting it right is important, though clearly not at all easy.

As we saw in chapter 4, object recognition is a difficult task for robots, especially in dynamic environments with variable or poor lighting conditions. Within buildings, one can supplement the robot's own sensing with smart building facilities.

This is not so easy in outdoor environments, especially for fast-moving vehicles. Hooking vehicles up to existing CCTV cameras would be one possibility but invites the question of what would happen if the vehicle drove into blind spots or otherwise out of camera coverage. This would also raise practical issues relating to data volume and ethical ones relating to privacy. As well as perception issues, there are anticipation issues. Humans use their superior object recognition to anticipate behavior—for example, that a bus will move from the bus stop shortly, but a delivery truck may stay where it is.[12]

The current state of the art in localization, navigating between waypoints and finding a route to a destination, does support autonomous cars. This has prompted the enthusiasm and the trials. The problem comes in the lower levels of autonomy, where systems have to make second-to-second decisions that can avoid—or cause—accidents, depending on how good they are. Train systems running on dedicated fixed rails are less risky as a result. There are also inspection applications that current technology would support well. AUVs for underwater pipelines, drones to inspect overhead electricity transmission lines, or even the state of domestic roofs are examples. Autonomous spy planes are also already in service with various militaries.

Indoor robot applications are in general more tractable but less attractive than autonomous vehicles from a business perspective. They break into specific applications, many of which can already be carried out easily and cheaply by people or by fixed networks, so that bootstrapping a mass market is hard. Many of the most desirable applications, for example, in health care or domestic support, require much more than navigation. In the next chapter, we look at whether we can let robots get really close to us—touching distance, in fact.

6

TOUCH AND HANDLING: COULD I SHAKE HANDS WITH A ROBOT?

The Deep Blue chess computer hit the headlines back in 1996, when it beat the world chess champion Garry Kasparov under normal championship rules—a triumph for artificial intelligence. Controversy ensued when Kasparov accused the IBM support team of cheating, and it is true that the researchers intervened between matches to modify the software against weaknesses the program had shown in the previous match. It is also true that Deep Blue was using its special-purpose hardware to consider millions of possible moves at a time rather than playing in any way comparable to a human approach. Nevertheless, although Deep Blue was retired and the hardware dismantled, it was the first in a series of world-champion-level chess programs such as the Deep Fritz program, which is at version 17 as of November 2019.[1]

So what has this to do with robots?

Deep Blue and its successors are brilliant at chess on the conceptual level, but the problem of moving real chess pieces on the board, a simple task for a human player, is outside their remit. The same is true of the AlphaGo program developed by Google DeepMind that beat the reigning Go champion in 2016. AlphaGo uses machine learning to good effect, but it does not move the pieces.[2] Human assistants do that. It's as if we do not consider motor tasks to be intelligent at all—very much in the Cartesian tradition of chapter 1, where only thought counts.

It is certainly possible to produce a robot arm to move chess pieces on the real board. You might think a standard industrial arm could do the job. However, chess has significant differences from most industrial applications. First, the objects are small, different from one another, irregularly shaped, and close together. Second, the arm does not know which piece to pick up or where to put it down until the next move in the game has been decided. Third, some of the pieces may vanish if the opponent takes them, and the arm has to remove opponent pieces it takes. Fourth, the game includes complex movement tasks like castling and a pawn being queened.

This combination of deciding on moves, vision processing, and precision manipulation makes chess a nontrivial task. This is why the Association for the Advancement of Artificial Intelligence (AAAI), the world's leading body for AI researchers, set playing chess as its 2010 and 2011 Small-Scale Manipulation Challenge, run at the association's annual conference.[3] Successful teams finessed some of the visual problems by changing the colors of the pieces to blue and yellow for better contrast and in some cases changing the colors of the board as well. They were asked to play ten moves from the standard start of a game. Thus the initial location of every piece was known. Until the final game in the 2011 run, each team played on a different board, with humans relaying their moves to their opponents. The challenge did not ask teams to solve the whole problem, which was regarded as too difficult for the challenge format.

If the task is only to move chess pieces, engineers can design a special-purpose gripper that works with chess pieces. A gantry over the chess board would solve the problem of pieces being close together. Good engineering can produce impressive results for a single task, especially if it involves only one arm and few objects in known locations. This is especially true for a robot arm fixed at its base. In this case, very little error occurs in localizing the gripper, since each arm joint has an accurate rotation encoder, and the length of each arm segment is known. Researchers in commercial kitchen automation are pushing to develop such arms,[4] and pancake and burger flipping are becoming feasible.[5]

You might see a theme emerging here: “special purpose.” Problems that have been reduced in size and scope to something very specific are generally soluble in robotics. When we try to achieve the generic capabilities humans and other animals have, we find that robots fall well short. This also holds true for artificial intelligence, as we will see in later chapters. Here, too, systems can solve specific problems and often produce better performance than a human. But the range of capabilities that humans have, sometimes referred to by the misleading term “general intelligence”—as with general manipulation—still falls well outside the state of the art despite some public hyping of systems.

Just as the game of chess is much more constrained than the real-world problems a robot faces, so manipulating chess pieces is still a simplified task compared to the general grasping capabilities we would like robots to have. The robot knows the number and shape of all the pieces; it knows exactly what manipulation is needed, so that this is a pick-and-place task. Making a cup of coffee in a domestic kitchen would be rather harder, and the robots that did well in the chess pieces competition would fail badly at making coffee unless the same level of design effort was involved. And this in turn would not enable the robot to load a dishwasher, a task set in 2010 for a mobile manipulation challenge at a robotics conference in Alaska.

Picking up a mug is a different problem from picking up chess pieces, especially if the mug is full of liquid. A gripper system that would be able to flexibly manipulate the same range of objects as a human would also have to recognize different objects and how to grasp them. This is the kind of knowledge a child acquires, starting at around six months, by grabbing a variety of objects and learning from successes and failures. We will see in chapter 8 that a whole branch of robotics, called developmental robotics, focuses on trying to mimic this infant learning process.

At the least, such a system has to recognize the affordances we discussed in chapter 2. Mug handles and wineglass stems are for grasping, lids are for lifting or unscrewing, depending partly on the shape of the object; some objects can be grasped with one hand, while others need two. A robot also needs knowledge of physical properties: containers may have liquid in them, in which case they cannot be tipped. Like the autonomous vehicle avoiding objects in its path, flexible manipulation in the everyday world really needs object recognition.

In an industrial setting, conditions can be tightly controlled. In everyday human environments, conditions—especially lighting—may vary all the time. Moreover, a really useful robot needs both the ability to move around in the environment and the ability to grasp objects within it. So instead of the robot arm being on a fixed base in a known position, it is mounted on a mobile robot base with the positional uncertainties we discussed in the last chapter.

Let's assume a robot has recognized an object, so it knows in principle how to grasp it. As with the simple approach to walking, if the position of the arm is known and the position of the object is known, then the joint movements needed to get the gripper to the object can be calculated in exactly the same way as for a leg, though with the advantage that overbalancing is unlikely. But how does the robot know that it has successfully grasped the object? If it has moved its grippers so that the object is between them—for a simple grasp—how can it be sure when to stop moving the grippers inward?

I can pick up the mug of tea near me on the kitchen table without knowing my exact position or that of the mug. My eyes combine with my proprioception to tell me before I start that the mug is within reach of my hand. Hand-eye coordination allows me to move my hand to the mug's handle, and my knowledge of how to grasp a handle makes me turn my hand sideways with my thumb and first two fingers in line with it. To perform the actual grasp, I move my thumb and two fingers so that the handle goes between the two fingers, resting on the second, so that all three digits are in contact with the handle, thumb above it (or this is one way of doing it). Just touching the handle will not give me a good grip, so I move my fingers until the force I am exerting through my muscles feels strong enough to lift the mug without it tipping. It helps that my fingers are cushioned enough to deform their shape slightly to that of the handle.

So part of the answer to how a gripper knows when to stop gripping is force feedback. Muscles feed force back by virtue of their structure: they are *compliant*, or flexible, and can compress or expand. A metal arm is not naturally compliant; it is a series of rigid metal sections with joints. However, including springs in the gripper and a sensor to measure torque can give it some compliance and a way of feeding back the force it is exerting on an object. This does, of course, make the gripper more difficult to engineer.

A different approach is to use a vacuum gripper, where the amount of suction tells the robot how much force is being exerted. This works well in an industrial context where the extra equipment needed to supply compressed air is not an issue, but its weight and complexity, and the power needed to drive it, make the vacuum gripper impractical in most mobile robots.

These limitations have led researchers to investigate materials that can produce effects more like human muscles. Pneumatic muscles offer one possibility, but we have just mentioned the problems with compressed air.[6] Engineers have focused on materials known as *electroactive polymers* (EAPs). These polymers change their size or shape when put into an electric field. They go right back to experiments in the 1880s with a strip of natural rubber. A hundred years later, in the 1980s, researchers discovered that combining polymers with metals in a composite material worked even better, with good changes of shape or size at one or two volts.

Textile-based solutions also exist, and another option is to use metal alloys that have a property known as *shape memory*. If nickel-titanium alloys are deformed when cold, they return to their original shape when heated, as if they remembered it. It is easy to heat metals with an electrical current, so these alloys can be used to produce movement without any other mechanical parts. Bundling thin wires made of shape alloys together produces a system that expands and contracts in the way muscles do. If you add a flexible cup on the end, this type of artificial muscle can even create an efficient vacuum by pulling it back without any need for compressed air.[7]

Artificial muscles are not as successful as human muscles just yet. In 2005 an international conference on electroactive polymer actuators and devices in San Diego, California, ran an arm-wrestling contest. Three robots with artificial muscles took on a human arm wrestler. The human beat all three easily. This technology is still new, but it could offer substantial benefits for robotics. And as we will see later, artificial muscles could be extremely useful for powered exoskeletons, as well.

So far we have focused on robots picking up objects. But what about human contact? Could a robot lift a person, shake hands, or even give someone a hug? The problem with the physical contact involved in these actions is that if anything goes wrong, the human could be hurt. We saw in chapter 2 that robot arms are quite capable of accidently killing people. So as with autonomous vehicles, robot arms need range-finding sensors that allow them to avoid obstacles. A circle of sensors around each of the arm segments should allow the arm to stop if it gets too close to something—or someone.

However, we do not want obstacle avoidance that works so well that the desired contact never happens. Imagine the problem if the robot needs to move its gripper into contact with an object, but an obstacle avoidance sensor insists this motion will produce a collision and halts it before contact can occur. The clash between the two objectives could leave the arm stuck or perhaps oscillating gently to and fro, continually trying to get to the object and continually pushing itself away. Autonomous vehicles would have the same problem if their obstacle avoidance prevented them from docking with a charging station. There comes a moment when obstacle avoidance must be turned off.

Obstacle avoidance should make the robot safe around humans, but we also need compliance in the gripper so that when it makes an intended physical contact with a human, it does so gently.[8] The simple approach is to drive the arm via a spring rather than directly through a motor, though the artificial muscles we just discussed would really help here. But how about giving the robot a sense of touch? We are familiar with touch-sensitive screens; could they be used on a robot?

The human sense of touch works through our largest organ, our skin, a complex three-layer structure of epidermis, dermis, and subcutaneous tissue. Embedded within it are multiple receptors concerned with pressure, vibration, texture, temperature, and pain. These are all connected by nerve fibers to the central nervous system, including the brain, so that our whole body is covered with a sensory system. Replicating the complexity of human skin is well beyond the state of the art today, but engineers are actively working on how to cover a robot with a simpler sensory surface—a sort of artificial skin.

The initial idea was a hard robot shell containing an array of range sensors, rather than just a ring around each arm segment. However, a soft skin has a number of advantages. A much greater range of sensors can be embedded, more as in the human equivalent, and it is safer for collisions, as well as offering a more pleasant tactile interface. Touch screens use pressure-sensitive electronic components—piezoelectric sensors—made of materials that produce an electric current when pressed. This current can be measured and action taken depending on the location. However, these sensors are metallic and not at all bendable.

Artificial skins use a range of plastics and other soft materials and can be stretched over any robot shape. This means that the embedded sensors also need to be stretchable and has led to the development of *soft sensors*. Conventional metal sensors need complicated linkages to deal with bending and stretching. New approaches use dense distributions of nanowires—functioning a bit like the hairs in human skin—in materials such as rubber composites, or photolithography to print extremely thin layers of metal onto the substrate.[9]

These skins are currently research prototypes, and though they give excellent touch resolution and sensitivity—higher than that of human skin—much more work is needed to integrate them successfully into complete robots. Like all the other sensors we have discussed, the data they provide has to be analyzed and used, and object—or surface—recognition is just as much of an issue. We will look at learning approaches to doing this in chapter 8.

This research focuses on making the technology work. But what is the possible impact of a robot skin on a human interacting with the robot? Japanese humanlike robots already have latex skin, though it is inert, without any sensors to detect touch. Flesh-colored plastic skins may evoke the "uncanny valley" reaction we talked about in chapter 2 because they are like a human skin at first glance, but not quite right at second glance. Human skin folds and wrinkles differently. It has more color variation owing to underlying blood vessels, not to mention flaws like moles and scars, and of course it is warm to the touch. An interesting piece of research adapted the robot tactile skin idea to a completely different purpose: an interface on a mobile phone.[10] The idea is that the phone could recognize gestures and even respond to being tickled. Sure enough, many people find this idea creepy.

Unlike robot vision or hearing, tactile sensors need to contact a surface to function. If we consider shaking hands, or even more so, hugging, we see that the person-to-robot contact is only the final element of the interaction. When we shake hands with each other, we make sophisticated decisions about exactly what to do when the hands are already in motion. Hugging is even more complex because it only works if one person hugs high and the other person hugs low. In both cases, this socially interactive contact is much more difficult than working out how to grasp a static object—and we already saw that this is extremely challenging for robot arms.

Up to now, all the robots we have looked at have been made of metal, gears, and motors. But why not a soft, squashy robot, a robot that can squeeze through gaps and wriggle around obstacles? A soft robot would also be much safer around humans. The development of soft sensors is part of a much larger and active field called *soft robotics*. The idea is to make robots entirely of compliant materials like rubber, polymers, or textiles.[11] Much work in this area is biologically inspired and, like the work on mobility we talked about in chapter 3, is often biomimetic. This means that it reapplies ideas from living creatures, usually invertebrates such as worms or octopuses, but sometimes plants as well. It involves totally rethinking many aspects of robotics, from materials to sensors to movement and control.

While the snake-inspired robots of chapter 3 were made of discrete segments, each controlled separately, a soft robot is more like a tube, controlled continuously, perhaps by internal fluid or air pressure or the sort of artificial muscles we talked about earlier. Remember the idea of *degrees of freedom* from chapter 3? Each degree of freedom is controlled with its own specific input, and we need six degrees of freedom to

reach any given point in a robot's working space. Adding an extra degree of freedom creates *redundancy*, so that the robot has more than one way of reaching each point. Soft robots do not have separate degrees of freedom; you might say they have an infinite number of degrees of freedom. This makes them *hyper-redundant*, so that they have an uncountable number of ways of reaching any point in their workspace.

This in turn makes soft robots difficult to control, one reason why they are still a research area. A robot made of metal can have stiff legs on which to walk or wheels on which to run, but soft robots might have to slither or wriggle. However, researchers at Yale University did recently come up with an innovative solution to moving a soft robot. They made a robot skin of elastic that included soft sensors and shape memory actuators. They demonstrated that you could wrap this skin around the legs of a stuffed toy animal and use it to move the legs so that it walked—or at least shuffled—along.[12] As with the human-looking robot skin, many people found this movement uncanny. The Yale researchers envision general-purpose skins that could be used to actuate all sorts of soft robot shapes.

Just as earlier we picked out a theme of special purpose versus general-purpose, we can see another theme here. Engineers can produce novel materials and complex mechanisms that in principle are more general-purpose, but this magnifies the problem of controlling them. You can see this in work on the development of robot hands.

Most robot grippers are quite unlike human hands. Robot grippers have two or three fingers that open and shut and a rotating wrist joint. Sometimes a robot has special tools rather than a gripper, or the gripper is made of compliant materials like rubber or polymers for special-purpose gripping. But if we put robots into everyday human environments, they will

face the challenge of replicating the things we do with our hands. So surely they need hands like ours?

As with legs, discussed in chapter 3, human hands are complex. Each of our hands has twenty-seven degrees of freedom—and remember each degree of freedom needs to be controlled. Each finger has four degrees of freedom; the thumb has five and works independently of the fingers. The wrist adds another six degrees of freedom. A widely available anthropomorphic robot hand has twenty directly actuated degrees of freedom plus another four actuated indirectly.[13] It also contains 129 sensors, including touch sensors on the fingertips, as well as position sensors on each joint, force sensing for each actuator, and other sensors for temperature and voltage.

This is great engineering, but think about adding two such hands to a mobile robot. How much numeric data will they generate? How fast could a robot process this data, along with other demands on its processing? How much heat is generated? How much power is used? Then the robot has to be able to plan grasping movements using the arm to which the hand is attached, as well as the hand itself, with all those degrees of freedom to control. Maybe we could teach the hand specific motions by motion capturing a human hand with sensors on its joints? Unlike the problem of actuating robot legs, at least here the robot will not fall over if we get it wrong. We can also imagine storing a human motion-captured grasp motion—for example, the coffee mug—and adjusting it slightly in action for mug handles of different sizes.

Although this hand is anthropomorphic, it is still not completely like a human hand, so we still need a transfer function to map the motion of a human hand onto it. The human hand can carry out movements that will not work on the robot hand. One research group recently tried to replicate the

human hand even more accurately.[14] They started by scanning the bones from the hand of a corpse. They point out that human joints are usually replaced by mechanical parts ("hinges, linkages, and gimbals") that do not function like human joints, especially the thumb. Then there are tendons and muscles. The hand this group has produced can be driven by a motion-captured human hand in real time and replicates many grasping actions. This work is pioneering but is unlikely to appear on a robot in the near future.

Outside the field of robotics, we can use much of this robot technology to develop devices for humans who have motion impairments, especially better prosthetics. Just as "robots will take over the world" is a sensational take on robotics, so the field of prosthetics is haunted by the idea that incorporating robotic technologies into the human body will transform humanity into something else. If the first group has mobilized around the concept of "the singularity" when, supposedly, robots will become "superior" to humans, then the second group has mobilized around "transhumanism," or the idea that technology can be used to transform humans physically and intellectually into some kind of super-cyborg species, "the bionic man." Somehow it always does seem to be "man."

Apart from the disturbing echoes of eugenics around this idea, it turns out that technological issues in prosthetics are as challenging as in robotics itself. Prosthetics technology is unlikely to support such ideas in the foreseeable future. Its focus is necessarily very much on supporting people who have lost usual human capabilities rather than on augmenting them; bluntly, we still have lots to do on this first objective. The fundamental issues of power, size, weight, and safety all have to be addressed, though new materials and power sources are producing good progress.

Progress is particularly noticeable in artificial legs—lower limb replacements. Applying ideas from robotics has revolutionized the design of artificial legs. Current models have carbon-fiber frames, hydraulic knee joints, and a built-in microprocessor for sensor data from the prosthesis. Specialist versions are good enough to support world-class athletes in competitions.[15]

These are *active* rather than passive devices, meaning they have a degree of autonomy. This may sound alarming, but remember the idea that autonomy can be layered. Automated stumble recovery has been shown to reduce falls by more than 60 percent in clinical trials, and that it makes users feel more secure is almost as important. Giving the leg small autonomous behaviors helps users undertake challenges such as climbing stairs one over one as do people without prostheses, rather than having to take each step separately.

Doctors have long known that amputees often reject their prosthesis; work in the United States suggests that up to half of all lower-limb amputees refuse to use their prosthesis, depending on what user group is surveyed, and between 10 percent and 20 percent of users abandon their prosthesis within a year.[16] Discomfort and the amount of effort required to walk are major factors. The new generation of lower-limb prostheses tackles both these problems. Cost and access are the major issues blocking widespread uptake.

As we saw, hands are much more complex, and arm prostheses are correspondingly less developed. Not only is the physical engineering more difficult and expensive, but it is not at all easy for the user to control multiple degrees of freedom, even in hands that are still typically far from anthropomorphic. Hand movements are much more varied than those of legs; legs are mostly about walking. The holy grail of prosthesis control is to find an easy and intuitive way of automatically reading the user's *intent*: the movement the user wants to carry out.

We have two ways of controlling a prosthesis. The more straightforward but less convenient is via controls on the prosthesis itself. A multijoint hand may have to be operated in this manner, with joints locked into position sequentially by buttons, really reducing its usability. At least these days the user can access the control facilities via a mobile app rather than by physical buttons. One current hand has gesture control, with four different grips that the user can select via a gesture with the prosthesis. Which four grips are available out of a larger repertoire can be changed using a mobile app.

The other approach is to interface the user's own body with the prosthesis. This is less cyborg-like than you might think. Surface electrodes can be attached to the muscles the user still has, close to the prosthesis. The electrodes can detect muscle activation, establishing that the user wants to move the limb. Unfortunately, we do not yet know how to decode exactly what muscle force or position is being signaled. Users can, however, be trained to consciously use these activations, so for a hand, they can change the grip selected for it.[17]

The most all-encompassing prosthesis challenge is the development of exoskeletons for people paralyzed by a broken spine. We saw in chapter 3 how important force feedback from the feet is to successful walking. People with a spinal break have no feeling in their feet and cannot judge where the ground is to walk on. There are, however, big potential health benefits in being able to stand upright and exercise, even if the actual forward motion is not very great. The standard approach is to strap people into a powered exoskeleton and give them a walking frame, as well, both for safety reasons and because the leg motion is driven by the user's arms. This is a long way from the vision of full mobility in an exoskeleton and does not help about half of the target population who are tetraplegic and cannot use their arms, either.[18] It is also exhausting for more than short-term use.

Recent work in France, still very much at the experimental stage, has demonstrated that a different way forward may be possible.[19] In this work, a user with tetraplegia had sensors implanted between brain and skin above the sensorimotor cortex, known to control movement. As with muscle signals, scientists don't know which signals do exactly what; but this user, with extraordinary patience, spent two years training the decoding algorithm that took the sensor data and turned them into motion commands. He did this in a graphical simulation with an avatar that he got to walk and touch various objects. Finally he tested the algorithm in the real exoskeleton—which was tethered to the ceiling for safety—and successfully walked nearly 150 meters in it, using the graphical simulation as a prompt. This approach has a lot of potential, but expense and effort are obvious barriers to widespread use.

One research area expanding faster than these medical applications deploys exoskeletons as aids to industrial manual labor. Some manual tasks are repetitive but are still too difficult for an autonomous robot. These may involve lifting heavy loads or keeping the body in tiring positions. The exoskeletons are designed to support the flexibility and superior judgment of a human worker with the strength and endurance of industrial robots.[20] An advantage of these industrial applications is that they involve limited mobility, which removes some of the safety worries.

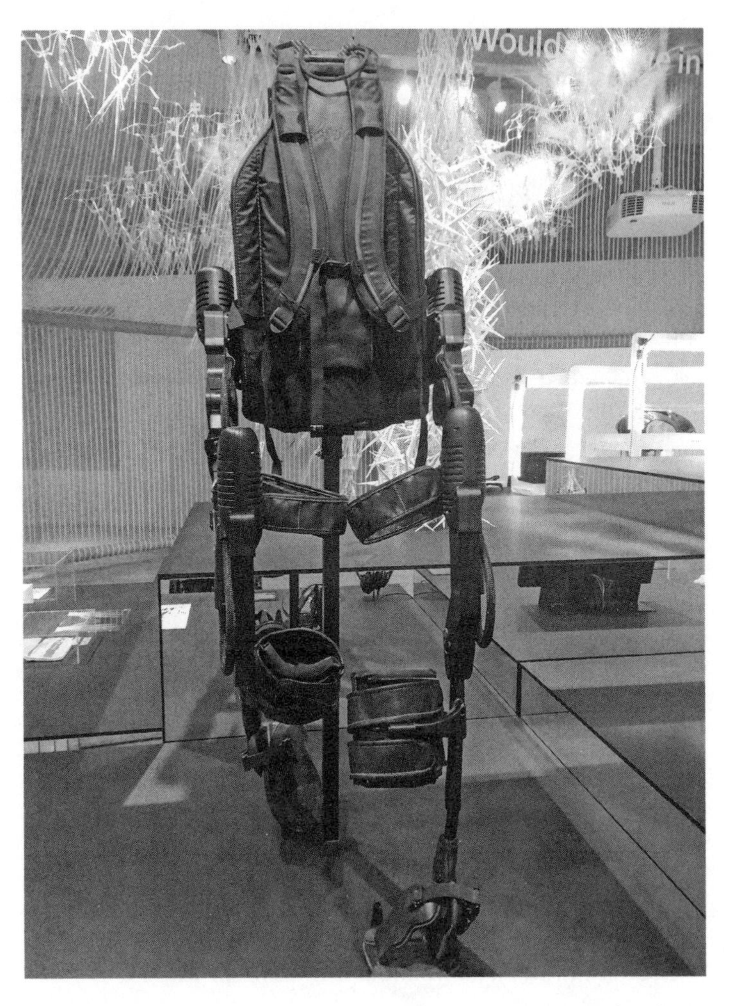

6.1 An exoskeleton to support a paraplegic. The arms are used to drive the legs forward.

Military exoskeleton applications are under trial, but this environment is much more demanding, involving uneven terrain and limited opportunities to recharge a powered exoskeleton. The US Army has tried a number of different designs, some of which are confined to bracing a soldier's legs and ankles. Both the US and Russian military are also testing more futuristic-looking full-body exoskeletons incorporating body armor. It is no surprise that these more complex constructions have substantial power problems, as we saw in chapter 3 with mobile robots. Running out of battery life in the middle of combat operations would be awkward.

Much less sensational, but also more widely deployed, are small-scale medical exoskeletons, often for arms and hands, used for poststroke rehabilitation. Poststroke rehabilitation exercises are repetitive and lengthy, and one way of engaging users is to link a partial exoskeleton to a computer game: a small-scale version of the graphical training in the French experiment mentioned earlier.[21] These exoskeletons are relatively inexpensive and largely driven by user movement within them, so they avoid the control difficulties we have been looking at.

What we see in this chapter is that though equipping a mobile robot with anthropomorphic hands and generalized manipulation capabilities is not anywhere near feasible today, specific spin-offs of these robot technologies are extremely useful. That is another message of this book.

7

COULD ROBOTS BE AIS?

A dog, an alien, and a robot walk into a bar, followed by a small girl.

"What can I get you?" asks the barman.

The dog gets up on its hind legs, leans over the bar, and selects a bag of chili roasted peanuts with its jaws and puts it on the bar.

"That's very clever," says the barman.

The alien points a purple tentacle at the bar TV, emits a blue glow, and "I'll have ten fluid ounces of a 3.5 percent alcohol beer" appears on the screen.

"Wow!" says the barman. "You're a really intelligent alien."

"I need to recharge urgent . . ." says the robot and collapses on the floor in a heap.

"Hmm, not so clever," says the barman.

He looks at the small girl. "Peanuts for you too?"

"No, thank you," says the girl politely. "But why did you patronize the dog? Why were you scared of the alien? And why were you so smug about the robot?"

Well, here's your starter for ten points. Which of these characters do you think is the most intelligent, and why?

Clue: there is no "right" answer. You can make a case for the intelligence of any of the characters (except the robot, of course). The dog is carrying out a sensible action in an environment not designed for it, the alien has solved the natural language translation problem (more or less), the barman is handling a complex social situation as well as dispensing drinks and snacks, and the girl has read the barman's nonverbal behavior to understand what he hasn't actually said.

Intelligence is very much one of Minsky's suitcase words; it holds a whole variety of meanings. In a single discussion, different people may mean different things altogether by the term. Dictionaries vary: the *Oxford English Dictionary* says, "the ability to acquire and apply knowledge and skills," and *Webster's* offers a veritable potpourri: "(1): the ability to learn or understand or to deal with new or trying situations; *also*: the skilled use of reason, (2): the ability to apply knowledge to manipulate one's environment or to think abstractly as measured by objective criteria (such as tests)."

Psychologists are more inclined to talk about a range of higher cortical skills: judgment, memory, learning ability, thinking, comprehension, orientation, calculation, language.[1] Then there is the concept of emotional intelligence—the ability to regulate and manage emotions—and social intelligence. Even the IQ tests referred to in the *Webster's* definition split the skills they claim to assess into areas such as problem-solving, short-term memory, spatial reasoning, and language. One could argue strongly that aggregating all these capabilities into a single attribute called "intelligence" has more to do with social rank than science and that really no such thing exists.[2]

So if intelligence is such a slippery concept, where does it leave artificial intelligence (AI)? How would we know if a robot had any, or whether it was "an AI"? How can we answer the question of the chapter title?

AI first became a research field in the 1950s, mainly in the United States. It was heavily influenced by the IQ test view of the world. This perspective is strongly connected to the Cartesian idea that intelligence is all about abstract reasoning and logic. The first AI system, the impressively titled General Problem Solver (of 1959), though intended as a model of human problem-solving, actually tackled brain-teaser-type problems.[3] Chess grand masters were seen as superintelligent, which explains the amount of research into chess-playing programs.

This view of AI saw it as allied to logic. How could a computer reason adequately unless it used a logic? Logics—there are many of them—are conceptually simple. They have a set of symbols of known truth or falsehood. Then they have some rules for transforming those symbols into another set of symbols for which truth or falsehood can be definitely deduced. Different logics work with different symbols and different rules. Logical symbols have no necessary connection to real-world objects, any more than integer numbers must refer to apples, oranges, or goats.

So it was quite feasible to implement a deductive logic in a computer and use it to solve problems that could be formulated in the logic. Games like chess can be represented in this way; chess has a small set of pieces and some simple rules for moving them. Mathematical theorems can also be represented in this way. But a stream of numbers coming from a robot sensor is not what a logic is intended for.

When we use symbols in everyday communication, we know what we mean by them. We ground our vocabularies in our own experience of the world. The idea of *redness* relates to seeing things that look red, and though integers do not necessarily relate to real-world objects, we learn to count by counting real things. How could a robot ground logic in its experience when its experience is numerical sensor data? We will come back to this issue in chapter 12 when we consider robots using language, our prime symbolic system.[4]

In the real world, humans need to process sensor data and act sensibly when it is not always clear what is true and what is false—not least because we use assumptions about the future that cannot be definitely proved. This may be why cognitive psychologists have found strong evidence that humans do not have deductive logic wired into their brains and do not necessarily use deduction to solve their problems. Identifying intelligence with deductive logic is problematic.[5]

Inductive logic is probably more useful to us much of the time. While deduction can prove true conclusions from true premises, inductive logic generalizes from the particular to the general and can fail. Since the sun has risen every day in our experience, we assume it will also rise tomorrow. Of course, it might not (the theme of some science fiction stories), just as our theory that all swans are white would be falsified when we came across a black one.

Some AI researchers ducked the problem of pinning intelligence down by defining AI as giving a computer the ability to do "things that are considered intelligent when humans do them."[6] This goes back to the application of skills in the *Oxford English Dictionary*'s definition. However, we should recognize that the skills we relate to intelligence are very much a matter of where and when we are. In seventeenth-century western Europe, the ability to do long division—dividing one multidigit number by another multidigit number—required a complex algorithm known only to a small intellectual elite. They were seen as really smart people.

Now that we teach long division in primary schools, and a computer can do millions of such calculations per second, we no longer see any arithmetic as requiring much intelligence. Overinfluenced by the Cartesian view, most people do not think complex motor actions require intelligence either, and object recognition—so hard for robots—is something we all do without thinking. AI researchers sometimes reflect gloomily that the minute their systems achieve a particular capability, that capability is no longer seen as "intelligent."

In the late 1980s, a section of AI researchers started out in a new direction. They took all the instances of "apply" in those two dictionary definitions much more seriously. They saw intelligence as a whole set of faculties and attributes that allowed humans—and other living things—to act competently in their own changing environments.

They also observed that apparently simple living things—bees, termites, cockroaches—managed successfully in their environments without anything like logic or reasoning capabilities. For these researchers, intelligence is always *situated* and *contingent*, so that it relates to specific contexts and to a particular history of acting in those contexts.[7] This definition emphasizes real-world interaction rather than reasoning over a knowledge base. This was very different from the timeless logic of the earlier approaches.

It is also an approach that is much closer to robotics. Everything we have looked at so far has concerned how to get robots to act successfully in the real world. So a useful definition of AI employed by many researchers today is the capacity to *do the right thing* at the right time in a specific context, where doing nothing would be worse.[8] Certainly the word "right" invites a lot of questions, but as an interactional, action-based definition, it does give us a handle on evaluating the "intelligence" of a robot. So maybe we should start by looking at what has to happen inside a robot for it to carry out any action.

We saw in chapters 4 and 5 that a robot has to work with the numbers it gets from its sensors. These numbers have to be turned into useful information about the robot's environment, for example, identifying a landmark it has seen before.

How does the robot interpret the numbers it is getting? Its software needs to know which sensor the numbers have come from: is it a video camera, a bumper, a laser range finder, an ultrasound transmitter and receiver, or something else? The type of sensor tells the robot what sort of information it is getting (pixels, range data, an impact, and so on), as well as how often (ten times a second, fifty times a second?). The specific make and type of the sensor will determine in exactly what form this information is encoded, how many bytes and what each byte contains, and how often it will arrive.

However, if a robot's sensor is replaced by a new one—maybe the old one has failed, or maybe there is a new one with better performance or a lower power requirement—this may change. To avoid having to rewrite all the software in the robot that is using data from a sensor, engineers write this software in layers. The idea is that the bottom layer is specific to the actual device on the robot, and then the next layer holds some standard numerical representation and translates the incoming data into that. Thus the robot has its own specific representation, and if the bottom layer changes, only the translation process has to be rewritten. Everything else can go on using the robot's own representation. This division into layers is part of the robot's *architecture*.

How much effort the robot puts into working on the sensor data depends a great deal on what the incoming information is. Say it came from a bump sensor that had been activated by a collision. Then without wasting any time on more processing, the robot should stop. In a human, we would call this a *reflex*. If you pick up a pan off the stove, and its handles are too hot, then your hands will let go of it. This does not require any thought on your part, but more than that, you have no choice: your hands will let go.

It is vital that a robot does have this sort of reflex for safety reasons. The first victim of a robot arm, Robert Williams, would not have been killed had the arm had a sensor that could activate a stop reflex just before impact. The autonomous vehicle that killed a pedestrian in Arizona should also have had a stop reflex.

There are other actions we carry out without really thinking about them. *Reactions* are not the same as reflexes; we have some control over them, and they do depend on the context we are in. When people greet you, you react by greeting them back, but exactly what you say or do depends on who the person is and where you are: you might hug a close friend or say, "Good morning, Madam," to your company CEO. Reactions require more processing of the incoming information than reflexes, but very little processing of the "what to do." We might think of reactions as being if-then sets of rules, and they can indeed be implemented as such in a robot. This was one of the earliest representations used in AI.

Using the term *rule* can be misleading, however. We usually think of rules in everyday language. For example: "If an unexpected obstacle appears in front of me, then avoid it." What this actually means in a robot is that a stream of range data is coming in, being translated, and being passed to a rule that changes the steering angle so that it gets bigger as the obstacle gets closer, and carries on being active until the obstacle has been passed. Some roboticists would call this a *behavior* rather than a rule, to capture the idea that it runs over a time period. The software layer that holds sets of these actions is the *reactive* layer.

As the steering angles are generated, they have to be passed back down to the robot's wheels. Just as with the incoming data, so the robot has its own representation, which is translated in its bottom software layer into the commands for the specific motors on its wheels. The bottom layer has more work to do, however. Remember the discussion of robot walking in chapter 3? We made a distinction there between kinematics—positional information—and dynamics, the forces at play, dealing with things like inertia and momentum. The reactive layer has decided the kinematics, but the bottom layer has to deal with the dynamics so that it can apply the right force to the wheels for the steering angle to operate correctly.

At this point you may be thinking: Where is the intelligence in any of this? Part of the answer is that the reactive layer of a robot normally holds a large set of reactions. How the robot will behave in any given situation will depend on which reactions get activated and for how long. Lots of sensor data is coming in all the time, so many behavioral rules may be active at once. How intelligently the robot behaves—how successfully it deals with its changing environment—will depend on how it manages all these calls on its resources.

Some behaviors can be combined. Obstacle avoidance movement can be added into movement toward a specific destination, so that if a robot is heading for a door and someone moves a chair into its way, the robot can move around the chair to reach the door. Other pieces of behavior may be incompatible: the robot cannot search a whole room for someone with a particular face at the same time as heading for the door, though if heading for the door is more important, the robot can look for the face on its way.

So how does a robot manage all its behavioral reactions? The researchers of the 1980s felt that the incoming sensor data could effectively manage the problem for them by activating whatever was needed at that moment. The dependencies between reactions could be handled by inserting links between them, forming a kind of network. These links could show which reactions conflicted and which were synergetic.[9]

If we add some internal sensors, then we can also give the robot behaviors, like one to search for a recharge station when its battery runs low. This reactive system can work well for robots that have a small number of things to do—for example, a vacuuming robot. Its big advantage is that the robot will always do something that relates to its current environment.

But what about a robot that has to carry out a set of very different activities, such as acting as a robot helper in a domestic environment? Building a huge network of behaviors with multiple connections is not easy to get right. What if two behaviors conflict for one activity but work well together in a different one? Testing, to show that it is going to work correctly, also poses a big challenge.

Earlier AI research had been based on giving robots the ability to make a *plan*.[10] The robot would build a model of its environment using sensor data. Then, using a database of possible actions, the robot would work out how to put the actions together in a logical sequence that should achieve its current goal. To execute the plan, each action would be expanded into all the actions needed to make sensors and motors work, and this would operate the robot.

We may think of planning as a characteristically human activity, a real mark of intelligence. In fact, some animals and birds are able to work out sequences of actions without there being much evidence that they plan the way humans do.[11] Furthermore, we "plan" a good deal less than you might think, usually only when confronting an unfamiliar task or when things go badly wrong. Most of the time we react and, when sequences of actions are needed, carry out a set that worked on previous occasions.

There are reasons for this. First, planning is hard work cognitively, and while we are planning, we have much less capacity for executing other actions. Sometimes we have no time to plan because we have to do something *now*. Second, to select the right actions for a plan, we need an accurate model of the environment. Sometimes this is hard to achieve, since we can only be really sure about what is in range of our own sensors. If we are planning to go out to the store for some milk, then we cannot know whether we will need to stop at traffic lights until we get to them.

On top of this, a plan relates to the future, so we must assume when we plan that the environment will not change in a way that frustrates us before all the planned actions can be executed. If I have just put my coat on, ready to go out to the store, and someone rings the doorbell with a package to be signed for, my plan has to be modified. So much of our planning activity is actually replanning.

These are all reasons why researchers in the 1980s preferred a sensor-driven approach, pointing out that robots using the earlier planning approach had worked very slowly, not noticed significant changes in the environment, and often failed part-way through a plan. We have seen in chapters 4 and 5 that a robot is unlikely to have an accurate model of its environment. In the real world, things frequently change, not least because of all the other people out there doing stuff. Avoiding unexpected change is one reason why factories must be designed so carefully for industrial robots. The original ideas around robot planning also gave the robot tunnel vision: it ignored everything except the actions it was supposed to carry out. If an unexpected hole opened up in front of it, a tunnel vision robot would fall straight in.

As with many battles of ideas, researchers began to wonder whether they could combine the responsiveness and interactivity of the sensor-driven approach with the goal-driven thinking ahead of the planning approach. After all, isn't that what humans do? We merge these approaches by using plans as a resource guiding us, rather than as a set of instructions detailing where each footstep must go. A plan tells us *what* we should do at some abstract level, and then as we go through it, we work out exactly *how* to do it given the conditions we find at the time. So my go-to-the-store plan could have an action "cross the road." But exactly how to cross the road would be determined by the state of the pedestrian lights when I get there.

Going back to the idea of a robot architecture with layers, we could add a third software level to the reflex and reactive levels already discussed. Rather than instructing the reactive layer in what order to activate reactions, a third layer could specify which reactions of the many available to the robot would be useful for the current goal. So to cross the road, the robot needs a reaction that will lead it to activate the pedestrian lights if nobody else has already, a sensor reaction that will be activated by the lights changing, and a set of navigation reactions that will get the robot to the other side. If the lights happen to be in the robot's favor when it arrives, then only the "cross the road" reactions will activate. We can think of the third layer as contextualizing the mass of reactions in the next layer down.

You may still feel that this isn't really intelligence, just putting actions together and then expanding them so they can evoke robot reactions that can eventually send numbers to wheel motors. How does the robot decide it needs to make a plan in the first place? How does it decide which goals it should try to achieve? Because researchers in the 1980s were interested in much simpler animals than humans, this wasn't really a problem for them. Their robots might be light seekers—carrying out phototaxis—that would switch to a recharge station when their batteries ran low.[12] Or they might explore their local environment and make a map of it, as in SLAM. Vacuuming robots are obvious descendants.

Even so, more complex robots that interact with people are still usually equipped with only a few goals. Carnegie Mellon University ran a cutting-edge years-long experiment with four CoBots—cooperative robots—that could carry out delivery tasks booked by users via a website, using similar scheduling software to Uber taxis.[13] The CoBots could deliver a variety of objects, including coffee, autonomously navigating a multifloor building with many similar-looking corridors and complicated spaces like a cafeteria.

These robots had a carrying basket but no arms; remember our discussion in chapter 6 about how tricky robot arms are to operate successfully. The basket was designed so that they could carry cups of coffee without spilling it all. The robots would ask humans for help as necessary—for calling the elevator, opening doors, putting the correct object in the carrying basket. This meant that their goals were largely confined to navigation, and the research around them involved developing robust localization methods so that they would know where they were in the building.

The CoBots show us why autonomous robots today have only a few goals. Every robot goal eventually gets transformed into movement of some kind. To be successful, it must stay within the set of movements the robot can actually execute, that is, map onto its set of reactions. The robot must have capabilities that equip it for the demands of the environment in which it is operating. Capabilities and environment jointly determine the goals a robot can execute successfully, without running out of battery partway. I would never adopt a goal for myself that required me to run fifty miles in a day, because I know I do not have that capability. Giving a CoBot without arms the goal of picking up a cup of coffee would be equally doomed.

We are inclined to underestimate the sheer amount of knowledge about the world we use when we make plans. A second group of AI researchers in the 1980s, not involved in robotics, focused on this area. For them, intelligence was less about having some amazing general problem-solving capability and more about knowing a lot, including how to apply existing knowledge to new situations. They started a field called *expert systems*, which tried to find ways of encoding knowledge about the world for use by computers.[14] However, just as robots work well within narrowly defined limits, so expert systems also worked well only as long as they were given specific tasks—for example, diagnosing bacterial infections using lab data or configuring computers from possible components.[15]

You might want to argue that nowadays robots have access to the internet, so surely they can have as much knowledge as they need about anything and everything? There are two big reasons why this is not the answer. First, the internet largely contains data, not information, and its data is not always correct, either (as in "fake news"). Not to mention all the sites devoted to pornography or gambling. It is the human users who turn this data into useful information. Even internet search algorithms, which do use AI technologies, deal badly with ambiguity: for example, if I search for "Capercaillie," I will get a lot of data about the bird, not the Scottish band of that name.

Second, internet data is presented largely in language form. Not only does this require more in the way of language skills than a robot generally has—as we will see in chapter 12—but there is nothing on how specific words could be transformed into robot reactions that would allow the robot to produce sensor or motor activity.

The idea that robots might be able to formulate their own goals sometimes worries people. We will see in the next chapter how far this is from being an issue when we look at learning. Right now, nearly all robots have only the goals their programmers give them. They are also given the high-level actions they will need for planning, and the mapping between those actions and the reactive behaviors that allow the robot to do things. These reactions also have to be programmed in most cases.[16] Anyone can learn how to program small, cheap robots, but you will find that, with few behaviors and limited mobility and sensing, there is a limit to what you can make them do.[17] They are fun, and a good way to learn programming skills, as well as to appreciate firsthand the difficulties involved. Robots of this kind do not have anywhere near the capabilities, either in hardware or in software, to behave in a way we would consider intelligent, even for the most skilled programmer.

We suggested earlier that a lot of human planning is actually replanning. A seldom-mentioned aspect of intelligent behavior is to notice when things have gone wrong and do something about it. To quote a popular saying, "Insanity is repeating the same mistakes and expecting different results."[18] Humans are not always good at this, and robots are very much less good than humans, as they are at most tasks related to sensing their context. Sometimes a robot is just not equipped to notice, as in a notorious vacuuming robot story.

Jesse Newton, in Little Rock, Arkansas, blogged that his cleaning robot was set to vacuum his living room at 1:30 a.m. each day so that it was clean when everyone got up. One night his puppy, Evie, had an accident on a rug during the night. The robot ran over it and distributed excrement all over the ground floor, producing, in his words, "a home that closely resembles a Jackson Pollock poop painting": twenty-five feet of poo tracks.[19]

Nor is this an isolated incident; the makers of Roomba, the best-known vacuuming robot, admit that it happens "a lot." If you think about it, a sensor that distinguishes between animal feces and dirt that should be cleaned up is not all that easy to engineer. A mass spectrometer would probably work, but not within the weight and price allowances of a small cleaning robot. Advice to owners of robot cleaners is to also clear away metal nails, screws, or plastic Lego pieces. Here we are back to object recognition again: a human cleaner would anticipate the failure that the vacuuming robot doesn't notice even after it has happened.

The trouble with equipping a robot to notice errors is that for many if not most tasks, things can go wrong in an extremely large number of ways. Limiting the number of possible errors is yet another reason for the careful engineering of factories that use industrial robots. Indeed, in logic, for any true statement, there are an infinite number of false statements.

My plan to go to the store could fail if there was a serious fire nearby and the area was cordoned off, or if the store ran out of milk, or if I tripped over a curbstone on the way and fell, breaking my wrist. As well as in many other ways. Some of these might wreck my overall goal of coming home with some milk, at least until my wrist was fixed. Some might require some rescheduling—the area will be uncordoned at some point. Others require a modified plan—if this store has run out, is there another one that will have milk?

Not only are humans better than robots at noticing that things have gone wrong, but they are also good at deciding whether to abandon a plan, modify it, or make a completely different plan. Maybe my neighbor can give me some milk if I really need it? We also have a good understanding of how important the goal was and how much effort to put into correcting the failure. Catching the train from Edinburgh to London would not be a justifiable use of resources to get the milk. A vital part of all engineering is considering possible failures and failing safely—so if a tall building collapses, it should do so downward and not by toppling over. Luckily, building construction has far fewer failure modes than for many of the tasks we might want a robot to carry out.

The simplest option for a robot, assuming it notices the failure, is to have it give up easily and report back for further instructions. Running over dog poo is a fatal error for a cleaning machine, and sitting there blinking a red light or beeping would certainly be the best action. This is also what planetary rovers do, on the grounds that trying something that also fails is risky when the rover cannot be rescued or mended.

All these examples reiterate our earlier point that we tend to underestimate the amount of knowledge we are using when we operate in the world. This is one of many arguments against the enterprise known as artificial general intelligence, or AGI, succeeding within any foreseeable timescale. This idea is a throwback to the General Problem Solver if it supposes that some generic mechanism exists that could produce whatever we think our suitcase phrase "generic intelligence" might be.

An AGI—or sometimes just an AI—is usually posed as a machine that can understand or learn any intellectual task that a human being can. This also takes us back to the Cartesian view of what constitutes intelligence. It is a concern for AI researchers, since today we see a great deal of publicity, usually from people who do not work in AI research, claiming that it is about to manifest. This publicity represents the third cycle of such overclaiming. The other two took place from the 1950s to the 1960s and then in the 1980s, followed on both occasions by an equal overreaction that relegated AI research to the fringes for an extended period; this is sometimes called "the AI winter."

We return to the AGI issue in the book's final chapter, when we look at the social implications of robotics. In the meantime, it should be clear from the discussion that we do not currently have any idea how to make a robot an A(G)I, assuming we can agree on what one would look like and what it ought to be able to do. We can, however, produce useful robots for specific purposes, and the scope of what we can do is increasing in interesting and sensible ways.

8

COULD ROBOTS LEARN TO DO THINGS FOR THEMSELVES?

In 1994, a graphics researcher named Karl Sims invented small graphical creatures he called Blockies.[1] Their name came from their construction; they were small random structures of linked 3D blocks, odd toy-town shapes. They looked nothing like real-world robots. But what Sims was interested in is extremely relevant to robots: he wanted to find out whether his Blockies could learn how to move on their own, rather than being programmed.

His problem was that a Blockie needs two things to make this work: a physical—or in this case graphical—structure that can be moved, and a control system to tell it when and how to do that. Sims was interested in mimicking the evolutionary process, which in the real world has produced living things well adapted to movement using their physical shape. We saw in chapter 3 that the physical mechanisms, sensors, and nervous system supports that allow humans to walk so well on only two legs are highly complex. So Sims made his Blockies much, much simpler.

He implemented a *genetic algorithm* that incorporates some aspects of evolution.[2] It represents a Blockie's structure as a set of numbers, and its control structure by another set of numbers. It generates a large number of different Blockies. Then it gets each control system to run its Blockie in a simulated physical world to see if it moves. Each is evaluated using a *fitness function,* in this case how far the Blockie managed to move. The ones that manage to move most are taken as the basis for a shuffle of the structure and control numbers to generate a new set of Blockies: reproduction of the fittest. Then the whole cycle repeats.

Sims found that with enough generations of this process, his odd structures could learn to "swim" (move in a simulation of a viscous liquid) and "crawl" (move across a graphical plain). When you watch videos of Blockies, you see a mixture of elegant and clunky. Their movement is not what a human designer would have programmed. Sims's repertoire of blocks did not include anything that would lead to wheels (which need axles and bearings), and the Blockies never evolved usable legs. But they did demonstrate a principle: there were processes that allowed simple structures to learn how to move, even if only in a virtual world.

Five or so years later, researchers at Brandeis University explored whether Sims's approach would work in the real world.[3] Rather than blocks, their shape repertoire was based on rods and joints. Using similar algorithms, the Brandeis team followed Sims's example and tested their evolved graphical creatures in a virtual world. But then they went a step further and output the successful designs to a 3D printer. Motors did have to be snapped in by hand, since they cannot be produced by current 3D printers. After this, the research team showed that their small rods-and-joints creatures could move in the real world as well as in the virtual one.[4]

We may not think of evolution as a learning mechanism, since in the real world it works over the timescales of many individuals, not just one. Isn't this adaptation, not learning? Well, if we define robot learning as "becoming able to do the right thing more often," then evolution does meet the definition. You may object that it is not how you think of learning. *Learning* is indeed another of those suitcase words that contains a set of different meanings. So let's try to unpack it a little.

The "what" of learning can vary widely. Learning the capital cities of European countries involves memorizing a set of symbols and associating them with "capital cities of Europe"—essentially a memory task. Learning which seedlings in the vegetable garden are weeds, and which are your own sown broccoli, is a visual pattern classification task. Other sensory classification tasks might include learning to distinguish one red wine from another or which instrument in an orchestra is playing in the music you are listening to.

Then there are motor skills: learning to walk, pick up random objects, ride a bicycle, swim. There is the learning of compound tasks involving memory, comprehension, and judgment, as well as senses and motor actions—for example, learning to talk, read, and write. But also how to safely drive a vehicle; how to solve mathematical problems; how to write novels; how to play the guitar; how to put up shelves. Not to mention social interaction tasks: how to conduct a multiperson conversation in different social contexts; how to teach; how to be a guest in someone's house. Not only does the content of learning vary, but so do learning processes. We can learn by rote, by instruction, by imitation, from experience, from imagination—and, of course, by mixtures of all of these.

Given the complexity of programming robots, having them learn instead is an attractive idea. Yet robots that learn may also seem a threatening prospect. Would the ability to learn take them outside of human control? What if they learned things that were antisocial or even destructive? Let's see if such fears are well-founded.

One characteristic of a robot is movement, and we have seen robot movement is generally not very competent. So being able to learn motor actions would be really useful. This is not an easy task, as the time it takes for a baby to become a competent manipulator of its own body shows. It is much harder than learning how to play a good move in games such as chess or Go. The physical world is much less bounded than a game and much more variable. Events in a game world are not ambiguous. Your opponent makes a well-defined move, and you must make a well-defined move back. In the real world, the robot is getting streams of numbers from its sensors, and it needs to know which are the right streams of numbers to send to its motors.

There are three main approaches to computerized learning: supervised learning, unsupervised learning, and reinforcement learning. What we actually mean by *learning* here is giving the robot an input and getting "the right" output where we didn't before. The idea is that if the robot covers enough input–output pairs while it is learning, then it will be able to produce outputs for inputs similar but not identical to the ones it has seen.

In supervised learning, we give the robot the input, and then we—as teacher—tell it what the output should be. In unsupervised learning, we give the robot a lot of data and leave the robot to find pairings. Reinforcement learning is a sort of halfway house: we give a robot an input, and then it generates an output and observes the outcome to see how good its response was—via a *reward* or a *punishment*. This is the most widely used approach today for research in robot learning, though it has yet to be applied outside of research.[5] Commercial robots use conventional handcrafted control systems, and there are some good reasons for this that we will come to. However, the popularity—not to say the hype—surrounding machine learning in other fields has given a big push to the idea of applying learning in robotics. Reinforcement learning (RL) is an attractively generic approach for those who would like intelligence to rely on one powerful mechanism.

RL is usually implemented using a network of all possible actions, where there is a probability value for passing down a link between one action and the next, a technique known as a *Markov decision process (MDP)*. RL algorithms look for the path through the network that maximizes the rewards, which it calls a *policy*. Robotics does have some key features that make it harder for RL than other domains.[6] One feature is that a robot trying out possible combinations in the real world may have safety implications; consider a "self-driving vehicle." It is also extremely time-consuming. RL algorithms can require many millions of examples to learn a good policy.[7] Consider just how many possible movements exist for a robot arm with six degrees of freedom. Battery limitations and possible mechanical failures are other considerations.

The Blockies learned in a graphical world, so maybe robots could learn in a simulator? This is an option pursued by much research into robot learning, but it has some serious limitations. A robot has real physics, and so does the world in which it operates. A graphical robot in a graphical world only has as much physics as whoever programs them chooses to implement. So the simulation would need to add noise to all the simulated robot's sensor readings, since we know this is an important factor in the real case. But what about variations in friction and other forces, motors not always working in exactly the same way, nothing being in exactly the same position twice running, and the many other imperfections and variations that make the real world such a challenge?

One of the authors once worked with two bin-sized robots called Fred and Ginger, which jointly carried a large object resting on sliding platforms on their heads. The idea was that if one robot moved a bit too fast, its carrying platform would slide back, while the slower robot would have its platform slide forward. The first one would slow down a bit, and the second one speed up a bit so as to recenter their platforms.[8] Think of two people carrying a table and using the pressure of the table on their hands to match speeds.

Both robots could avoid obstacles, and the researchers noticed that in their simulation, if the two robots got to an obstacle like a pillar exactly side on, then one robot might go left and the other right, with fatal consequences for their carried object. But when the real robots were used, this event never occurred, because no two real robots are actually identical. One robot would always have a bit more battery power, or a bit less wheel slip, and the faster robot would pull the other one around the obstacle just fine.

On the other hand, the biggest problem with the setup never appeared in the simulation at all. If one robot speeds up and its sliding platform goes back, the other can speed up slightly too much, and the robot pair can start oscillating wildly, with the changes in one increasing the changes in the other. This effect is due to the dynamics of the system, and it is no surprise that it did not occur in the much more well-behaved simulation. In summary: the problem that at first looked most serious only ever appeared in the simulation, and the problem that actually *was* the most serious never appeared in the simulation.

Groups researching RL with robots are aware of these issues, and of the need to use simulations cautiously, perhaps only to get robot behavior into the right range. That something works in simulation does not mean it will work with a real robot. The first question one should ask about impressive-looking results is whether they happened in simulation or in the real world.

One alternative involves learning from a human performing the task. You need the actions that the human is performing—which must be ones a robot can perform in exactly the same way—and you also need sensor data for the outcome of the actions, so that the system gets the reward input. These are both available where a robot automates a driving task and you can record a human driver. Recent work at the University of Cambridge used RL to have an autonomous vehicle learn lane following by using eleven videos of a human driving.[9]

The amount of experience a real-world robot may need to learn using RL is not the only difficulty in implementing it. Remember the reward function? This is a critical part of the process, since it is effectively telling the robot what to learn, just as the fitness function told the Blockies how they were doing as they evolved. Designing a good reward function is not straightforward. A reward function needs to be able to return a result often enough during a task to guide it, preferably after every action. One reward right at the end is not so useful. The reward also needs to be numerical, a matter of degree, rather than all-or-nothing; a binary success/fail is much less useful.

In the lane-following example, the researchers tried a reward of driving as far as possible without safety driver intervention. This sounds sensible. They discovered that the autonomous vehicle learned to zigzag down the road. It did not leave the lane and cause a safety driver intervention, but by zigzagging drove a greater distance, maximizing the reward. Another researcher cited using RL to teach a (simulated) robot to stack one graphical block on another.[10] The reward was initially set to be any increase in the height off the ground of the bottom of the block to be moved. This will happen, they reasoned, if the arm lifts the block into the air. Unfortunately you can achieve the same thing with less effort by tipping the block over. Researchers in the field call these unintended behaviors *reward hacking*.

Some alarmists have extrapolated these issues to fictitious robots that might then cause major problems. The "paper clips" thought experiment poses the idea of a robot with the single goal of maximizing its output of paper clips that might then create a policy annexing all the earth's resources, including all human bodies, as potential raw materials.[11] This is fine as a thought experiment, but anyone struggling with RL applied to a two-joint simulated robot stacking one graphical block on another would point out that it is as likely as hell freezing over.

As a thought experiment, it is also imprecise: Does it mean a robot that has learned to bend a wire into the right shape? A robot that can make the wires as well? A robot that has learned to mine the correct ore, turn it into pure metal, then learned how to transport goods and how to deliver them so as to make the wires? A robot that has learned to extract metal from human bodies? And so on. It does, however, underline a significant point: a robot with just one goal is unlikely to be beneficial to the world at large. Human monomaniacs are perhaps a more immediate danger, however.

An autonomous vehicle needs a whole set of rewards and punishments. Its long-term goal is to get its passengers safely and quickly to their destination. Important additional goals are not to injure (or anger) other road users, break rules of the road, or cause traffic accidents. Not all of these are easy to turn into numbers. In the real world, reward functions are necessarily multifaceted and may involve trade-offs. Any useful robot has to learn more than one thing: lane following is one of many skills needed for an autonomous vehicle. That learning new skills can result in losing good policies for older ones is also an RL issue. These are all reasons why RL learning of robot motor skills is still a research area.

One thing we do know is that humans are effective learners. So why not construct the control system of a robot to look like a human brain? Wouldn't that make a robot as good a learner?

When people talk about "robot brains," they are usually referring to a structure called an artificial neural network, or ANN. This is a computer-based model drawing on some of what we know about the structure of human brains. The human brain contains about eighty-six million neurons, each connected to as many as one thousand others. A neuron receives inputs via feathery, treelike connections called dendrites, carries out complex electrochemical processing, and sends its output, as an electrical pulse, down a single structure called an axon, which ends at a synapse. The synapse connects to other neurons via their dendrites, and a typical neuron sends out pulses between five and fifty times per second.

Researchers worked on modeling networks of neurons way back in the 1940s, well before digital computers were invented. The basic idea was a data structure representing an artificial neuron, with an algorithm to weight and then sum the inputs it received, and if these were greater than some threshold, send a pulse on in its turn. When artificial neurons were connected together in a network, different patterns of inputs to neurons could be converted into outputs from neurons they connected to. Learning could be modeled by altering the weights of inputs: upward to reinforce better answers and downward to inhibit less good answers. The weight changes would produce a different pattern of firings and thus different outputs, and the whole process could repeat until the outputs were right.

ANNs had a roller-coaster ride as researchers were in turn enthusiastic about, and then disappointed by, their capabilities. Many of the techniques we call "AI" can also be thought of as mathematics. Mapping inputs to outputs is what mathematical functions do. Drawing the best line through a lot of x,y points, where x is an input and y an output, is fitting a function to data using regression, which is a statistical technique. In the 1960s, a wave of disappointment set in as it turned out that some quite straightforward functions could not be computed by the simple networks then being used, which had only an input layer of neurons and an output layer.[12]

Other researchers suggested that inserting extra layers could increase the scope of ANNs, but for a long time, nobody could work out a sensible way to adjust the weights on the neurons in the middle of a multilayer network to model learning. In the mid-1980s, a different algorithm, *backpropagation*, was popularized for doing this.[13] The idea was to feed in some inputs, see what errors this produced in the final outputs, and feed the error backward through the network to adjust the hidden weights. By cycling forward and then backward enough times, the network would usually settle down and then become capable of correctly processing inputs it had not seen before.

This produced another wave of enthusiasm as the new ANNs were applied successfully to tasks like recognizing handwritten letters and digits, where more conventional methods had never produced good results. These ANNs were computationally ingenious and practically useful but started to diverge from what was known about the brain. No evidence indicated that human neurons did backpropagation. Some researchers became highly optimistic that ANNs could solve all the problems that AI was still struggling with; others warned that ANNs too would turn out to have limitations.[14]

Backpropagation is a form of supervised learning. You have to tell the network what the outputs should have been to calculate the errors it backpropagates. Alongside these networks, others were devised that behaved like statistical classifiers. These are able to cluster together inputs that are "similar" in the same range of attributes, so that they are close together in some multidimensional space. This is a type of unsupervised learning. Finally, ANNs that dealt with time-dependent data like speech came into play—recurrent ANNs in which outputs were fed back in again as new inputs.

In the mid-1990s, another hiatus set in as researchers found their backpropagation algorithm became ineffective where many hidden layers were present—a *deep* ANN—or the ANN was recurrent. Solving this problem led to the current renewal of enthusiasm. Whenever you hear the keyword *deep* alongside machine learning, you are hearing about ANNs. We have seen many successful applications from the 1990s onward—though remember that once AI technology comes into common use, it usually loses the AI name. However, most of these applications were in sensor-processing and classification tasks, not in control. That made them useful for robot sensing but less so for driving robots. There was a good reason for this.

Engineers who write control systems, whether for robots or other equipment, need to know how good they are. Under what conditions do they function correctly? Are there conditions under which they do not? Are they stable, or could they be susceptible to that runaway feedback that the Fred and Ginger robots discussed earlier sometimes suffered? Engineers have a battery of mathematical tools for answering these questions with conventional control systems. Even when things go badly wrong, as they recently did with a Boeing fly-by-wire system on aircraft, such tools can identify the error and thus how it can be fixed.

Analyzing a control system written as an ANN is problematic. The way the control system works is determined by the weights in the ANN and by the data on which the ANN was trained to produce those weights. Though an ANN is a mathematical function, it is not possible to retrieve exactly which function it implements, since this is distributed throughout the network. It means you cannot know exactly what a deep learning system has learned: its workings are opaque, not transparent.

Though deep learning has been successful in image-processing problems, often working better than humans, some unfortunate errors have demonstrated how sensitive a deep learning system can be to its training data. Notoriously,

Google Photos, which uses deep learning to label photos, offensively labeled pictures of people with dark skins as gorillas, possibly because the system had an insufficiently varied training set.[15] Even successful classification systems may not always be using the features people expect; in 2018 an article raising questions about the scope of deep learning quoted a paper saying that the "highly accurate discrimination of wolves from dogs on a dataset extracted from ImageNet was the result of detecting white snow patches in the wolf images."[16]

These limitations have not stopped researchers working away at the issue of applying deep RL to robot motor skills. One of the nonprofit companies working on AI has demonstrated a robot hand, with twenty-four joints, manipulating a Rubik's Cube.[17] Remember, we saw in chapter 6 how hard it is to successfully control a robot hand with many joints. The company's video shows the robot hand manipulating the cube one-handed, starting from different initial states, and carrying on even when hit with a soft toy, a stuffed giraffe. While robots specially engineered just for this one problem can solve it much faster than this robot (and faster than a human), learning a motor task of this complexity is most impressive.

As critical reviews have pointed out, however, this is still a long way from solving generic problems of grasp.[18] The robot could not pick the cube off the table, dropped it at some point in 80 percent of the trials, and had to use a cube with special sensors so that the robot could tell what state all the faces were in. Remember too that RL usually requires a large number of trials. One of the interesting research aspects of this robot was the reasonably successful transfer of learning in simulation to the real world, using random disruptions in the graphical world to increase the robustness of the movement in the real world. However, reports suggest that it took the simulation equivalent of ten thousand years worth of human manipulation for the RL to succeed, involving immense amounts of computer power.

The AI revolution of the 1980s that we discussed in chapter 7 eventually produced systems that tried to merge the new data-driven approach with older model-based ones. Some researchers are now exploring whether the opaque but generic methods being used in machine learning could be combined with more explicit knowledge about a particular domain to improve performance and scope. They hope to reduce the extraordinarily large number of trials RL requires, as in the Rubik's Cube robot. One piece of recent work involved a robot learning how to play the game Jenga, which involves a tower of wooden blocks.[19] The idea of the game is for each player in turn to remove a block and put it on top of the tower without collapsing the whole structure. The Jenga domain has attracted many researchers interested in robot manipulation.

The robot used in this research was a standard industrial robot arm, but with added sensors: a force sensor on its wrist and a separate camera trained on the tower. It had a two-pronged gripper, much simpler than the hand of the Rubik's Cube robot, and was programmed to explore the physics of the tower by prodding blocks at random to see what the results were. It used both its sensors to assess results, combining a measure of how much force it had used with a visual record of what happened.

Rather than only using models, or only using data, this project combined both in a hierarchical approach. Abstractions at the top of the system represented knowledge of the physics, and data at the bottom represented the outcome of actual moves. Because the model guided the way the robot got its data, it was far quicker than the flat try-anything approach of RL. It was also closer to human learning, where we see children experimenting with their environment and gaining a general idea of everyday physics.

Exploring how children progress from helpless babies to functional adults—known as ontogeny—gave rise to yet another approach to robotics, named *developmental robotics*. This area is relatively new and seeks to develop robot capabilities through the intersection of developmental sciences, cognitive science, and robotics. The idea is twofold: to use robots to validate developmental science models, and to create better robots by uncovering the intricacies of ontogenetic development.

Developmental robotics overlaps with epigenetic robotics, a similar field of research that also looks at cognitive and social development plus interactions between sensors, motors (sensorimotor interactions), and the environment. However, developmental robotics goes beyond epigenetic robotics by investigating the links between motor skills and morphological development.

Many developmental robotics platforms have been developed to date. The most famous is the iCub robot. The iCub is a humanoid robot developed at the Istituto Italiano di Tecnologia and is used today in more than twenty research laboratories.[20] Because its design is based on a human child, this humanoid robot helps researchers to better understand the innate development of infants, and the power that drives this development plus the inherent acquisition of knowledge that accompanies it.

One research group recorded the trajectories of the limbs of real crawling babies.[21] Based on this data, they designed an oscillator driving a pattern generator network. This model

8.1 The iCub robot was developed specifically to support research in developmental robotics and is modeled on a child.

reproduced almost the same motion in a dynamic simulation and was then ported into the iCub.

Developmental robotics is an emerging field and has branched out from other theoretical frameworks like "autonomous mental development," "evolutionary developmental robotics," and "cognitive developmental robotics."[22] Work in the area includes modeling the development of child motor skills, as in the crawling example, but also hand-eye coordination, turn taking (as in peekaboo games), and the development of joint attention.

"We called him Tortoise because he taught us . . ." marks the start of a seminal paper by Grey Walter about his "mechanical tortoises" in the 1950s, which most believe is the first neuro approach to robotics. Grey Walter was born in the United States but grew up in the United Kingdom, where he spent his career investigating neurophysiology. As part of this work, he built small robots in the late 1940s, using the valve technology of the day. Domed and wheeled, they did look rather like mechanical tortoises.

Grey Walter's robots were simple mechanical devices that had only two sensory organs and two electronic nerve cells but could exhibit interesting and complex behaviors using attraction and repulsion behaviors like the phototaxis discussed in chapter 4. Grey Walter was trying to demonstrate that complex behavior could be produced via connections between quite simple elements alongside interaction with an environment. One of his original tortoises was put on display in the Science Museum in London in 2000.[23]

In a quest for ways to create better robots, some roboticists are looking for further inspiration from human brain architecture, brain dynamics, and neuroscience findings. To this end, more realistic models of neurons, first proposed in the 1950s by Nobel Prize winners Alan Lloyd Hodgkin and Andrew Fielding Huxley, are now being revisited. The neurons of the ANNs described earlier are mostly approximators of nonlinear but continuous functions. They are processed synchronously, updating in step with one another on the same clock beat. Biological neurons like the Hodgkin-Huxley model can fire in asynchronous spikes, responding to events from elsewhere at any moment. These *spiking neurons* are being used to create artificial models of regions of the brain, simulating what we know of how these regions work, and then embedding these models into a robot. This field of research is called *neurorobotics*.[24]

Neurorobotics is a relatively new field of research, though it was influenced by the sensor-driven, behavior-based robotics approach of the 1980s and 1990s. Brain-body-environment coupling is its central concept. Its main principles for creating intelligent systems or robots include changing the way robots learn, by using more biologically plausible neural models like spiking neurons, and simply rewiring artificial neuron connections. Neurorobotics embeds artificial brain regions into a robot body because real-world interaction depends on embodiment. It designs more flexible representations to better deal with environmental constraints.

One interesting example of a project in this field is looking at neurodegenerative brain diseases like Parkinson's, trying to design more biologically plausible models of the brain regions involved. By running these models on a robot to see what physical symptoms result, scientists can check the model and then reapply it to our understanding of the human case.[25]

In this chapter, we have looked at different ways in which researchers are trying to develop intelligent robot controllers through learning processes rather than through being directly programmed. The Rubik's Cube robot shows that individual sensorimotor tasks can be learned, though slowly and with immense effort. However, we have yet to see a robot capable of lifetime learning. General learning is as far from robots today as general intelligence or general grasping abilities.

9

COLLABORATING ROBOTS: COULD THEY WORK AS PARTNERS OR GROUPS?

Have you ever played Conway's Game of Life? Not the real-world game where you get up and go to work, or even the board game, but the computerized one invented by the British mathematician John Conway in 1970?[1] The game plays out on a 2D grid of empty squares. You fill in a few at the start, each of your squares representing a tiny creature (strictly speaking, a *cellular automaton*). Once you have set up your population, the computer runs a set of simple rules:

1. Any live cell with fewer than two live neighbors dies, as if by underpopulation.
2. Any live cell with two or three live neighbors lives on to the next generation.
3. Any live cell with more than three live neighbors dies, as if by overpopulation.
4. Any dead cell with exactly three live neighbors becomes a live cell, as if by reproduction.

Some starting patterns die out quickly. Some simple ones continue indefinitely, like the Block, a square containing four live cells. Some oscillate, like the Blinker, a bar of three squares, alternately horizontal and vertical. But the interesting thing is that the game can create and then reproduce patterns that nobody has programmed. Take the Glider Gun, a circular structure that produces a pattern every so many cycles that moves off across the board like a bullet.

These patterns are examples of *emergence*, the creation of complex structures or processes from interaction between much simpler ones.[2] Weather systems are one of many examples in the natural world, arising from interaction between molecules of air and water vapor. There are strong arguments that human consciousness is an emergent phenomenon of interaction between brain structures.[3] Other examples involve living creatures, in particular social insects like ants, termites, bees, and wasps. These insects have fascinated some roboticists.[4]

Social insects are capable of large-scale collective activity without any centralized direction. Termites build extraordinarily complex structures. One species creates mounds as large as thirty meters across and six meters high. Another species creates external structures that work as air-conditioning, and connects internal chambers with ramps. Ants will quickly focus collective effort on a new food source found by one ant. Estimates suggest an individual ant has maybe twenty basic behaviors—more than in the Game of Life, but not on the scale of mammals, for example.

We have seen how complex one big robot can get. Could we use the idea of emergence to create complex behavior from the interaction between many simple robots instead? This approach has possible advantages. Because each robot would be simple, building many robots would be less expensive; and if one robot failed for some reason, then the task could still be completed by others—redundancy is built into the system. You could add extra robots, if available, without reorganizing everything. A set of robots might also be more flexible, able to adapt to changes in tasks. Their software ought to be much simpler than a large multipurpose robot.

As with the biomimetic applications we looked at in chapter 3, this approach involves investigating how social insects manage to accomplish complex tasks. The first mechanism uncovered was the communication of information via the environment, known as *stigmergy*. A human example is the formation of a footpath by people walking where the ground shows that other people have been walking, called *desire paths*.[5] Researchers found that wasps built their nests, constructed with chewed up wood pulp, by feeling for a building site with their antennae. They tended to go for a site at a corner area where three adjacent walls were already present. Termites also used the configuration of the structure to add to it, and as they added to it, the configuration changed. Simulations showed that this local information was enough to produce the nest structure.[6]

Ant foraging using stigmergy has become particularly well-known in the computing world.[7] An ant that finds a good food source collects some and takes the shortest route back to the nest, laying a trail of pheromones. An ant that comes across a pheromone trail follows it, collects some food, and lays more pheromones. Pretty soon a big pheromone arrow is pointing at the food, recruiting lots of ants. If ants hit an obstacle, some may go one way and some the other, but the shorter route will get a higher concentration of pheromones as ants traverse it more quickly. When the food is exhausted, no more pheromones are laid, and the existing trail decays. Where the wasps were reacting individually to the part of the nest they were constructing, the ants are collectively self-organizing.

Drawing on these ideas has led to the research area called swarm robotics, which looks at how groups of at least a hundred small robots can be mobilized. One branch of this research explores developing the robots themselves and how to make them as small and as cheap as possible while still collectively proving useful services. Scientists have developed at least one experimental setup of more than a thousand robots.[8] Unlike social insects, these robots do not usually have multiple (or any) legs, since, as we saw in chapter 3, legs require more motors and thus consume more power. Miniaturization and the use of simple sensors are both important.

Another element of this research looks at applications where swarms would be useful. Dangerous environments are one target, since they might result in robot damage that a swarm would cope with much better than a larger single robot. These might include underwater or extraplanetary environments, as well as minefields and buildings wrecked by earthquakes. Distributed sensing tasks are another good domain, for example, monitoring oil spills or other ecological disasters, or even moving inside the pipework of process plants. Some of these tasks do not require collaboration, though they could be completed more efficiently and robustly by swarms. Others cannot be completed at all by a single robot, for example, materials transport. This is why a standard experimental domain is collective transportation, rendered in labs as box pushing.[9]

Taking miniaturization right down to the nano level—components at or close to 10^{-9} meters, or a billionth of a meter—would produce nanorobots, an area that has attracted a good deal of industrial interest. The metal-and-motors paradigm that dominates robotics, the basis for nearly all the discussion in this book, is not usable at nano scales. The idea of nanorobots involves creating mechanisms engineered from natural molecules, using biochemical processes for movement and actuation. Nanorobotics is a branch of biophysical engineering rather than a branch of robotics. It forms a small part of the field of nanotechnology, which was subject to its own wave of hype and public panic in the recent past. The motivating idea is to use swarms of nanorobots inside the human body to deliver medicines precisely where they are needed, and perhaps to deal with partial blockages in the vascular system before they cause a major problem.

Though this research has gathered substantial industrial investment, much of the work until now has gone into developing basic mechanisms. This is a long way from the more grandiose visions. What does a molecular motor look like, and how well does it work? Where does its energy come from? How would a nanorobot sense its environment? How does it navigate? Recent work has begun to demonstrate some useful results, with in vitro demonstrations of a nanorobot that can deliver a payload to destroy cancer cells.[10] Moving from in vitro to in vivo is a long process, however, and we should not expect any breakthroughs to happen quickly, especially given safety and ethical concerns.

The natural-world phenomenon that is now widely implemented in swarm robots is *flocking*: the ability of birds, fish, and some other animals to move in a formation. Think of the V shape of migrating geese or starling murmurations. Like the social insects, these animals have no central direction, and though birds are more complex creatures than ants or bees, they are not language users and do not agree on a plan before they move off.

As with the Blockies of chapter 8, flocking was initially investigated by a graphics researcher, rather than a roboticist, and has had a substantial impact on film animation.[11] This work showed that flocking could be simulated by applying three rules to each of a set of small graphical entities known as Boids (as distinct from birds). Each Boid needed to move so as to do the following:

1. Keep outside some threshold value between itself and neighboring Boids (separation).
2. Not allow itself to get too far from the Boids around it (cohesion).
3. Steer in the average direction of the Boids around it (alignment).

These rules differ from those in the Game of Life and are usually applied in 3D, not 2D, but the similarity is clear.

This work interested companies wanting to produce satellite formations in earth orbit, as well as those with undersea applications. The recent availability of cheap airborne drones has expanded access to many other researchers, and military interest is a given. Being able to use GPS simplifies localization, and applications that are being investigated include—inevitably—surveillance, as well as tracking chemical plumes and other environmental tasks. On the lighter side, Intel produced the Shooting Star drone for light shows, equipped with LEDs. These have been used in large numbers for ceremonies at public events. The biggest to date employed 1,218 of them at the opening ceremony of the 2018 Winter Olympics.

Before we run away with the idea that the world will be crawling with small robots in the near future, it is worth pointing out that we have yet to solve many problems before

9.1 Cheap drones support experiments with formation flying.

swarms can be applied for real. Swarms of robots are as subject to power constraints as larger ones. Moreover, being small, they are affected more by the low power densities of batteries, which are heavy for the power they hold. Many lab demos take place on powered tables that allow the swarm robots to continuously recharge.

More conceptually tricky is how to deal with the engineering challenges of emergent behaviors, described as *swarm engineering* by one leading researcher.[12] How to establish what behaviors swarm members need to produce a desired emergent behavior? How to be sure the emergent behavior will emerge correctly and under the expected conditions? How can the engineering requirements of dependability, fail-safe, and diagnosis of faults be met? You might note that these questions are similar to those raised by robots that learn, discussed in chapter 8. Mathematical modeling and formal verification approaches are being developed, but engineers still have much to do before they can be confident about a swarm approach in real-world applications.

So far we have talked about *collaborating* robots. But we have two other words for collective activity: *cooperation* and *coordination*. These words are not merely synonyms; they have subtly different meanings. In collaboration, entities devote their activity to a common vision, as when ants collect a food supply. In cooperation, entities support one another's goals, which are not necessarily identical. We can think of cooperation as a looser form of organization. Coordination is the mechanism through which collaboration or cooperation is accomplished, so stigmergy is a coordination mechanism for ants.

Where collaborative robots can be thought of as a swarm, cooperating robots can be thought of as a team. Human teams are everywhere: in offices, dealing with emergencies, in sport, in the military, on construction sites, in manufacturing. If we want robots to fit into ordinary human environments rather than remaining in specially engineered factories, then being able to deal with teamwork is a requirement. So how about purely robot teams? Remember the Carnegie Mellon CoBots of chapter 7? Four of them were available for various fetch-and-carry tasks in one of the university departments, and which CoBot would appear for which job was automatically decided by a scheduling algorithm. This is how they were coordinated.

The biggest test bed for all-robot teams does not necessarily have central coordination—it is robot soccer. There are two international leagues for robot soccer, RoboCup and the Federation of International Robosports Association (FIRA). RoboCup was set up back in 1996 by a group of computing professors who wanted to stimulate AI-based robotics with a standard but fun application domain.[13] It was originally posed as a "grand challenge" of producing a robot soccer team that could play against a human team by 2050, but quickly diversified into a whole set of subleagues using different sizes and types of robot platforms, as well as a simulation-only league.

Robot soccer is challenging because the environment is dynamic; in fact, it changes very quickly. Robots must make decisions in real time, processing uncertain sensor data, deciding what to do and then doing it, and they cannot just react. They must anticipate not only where the ball is going but also what their teammates are going to do and, as importantly, what their opponents are going to do. Anticipation as a basis for coordinated action is required for successful teamwork in most fields.

Everything we have discussed so far about humanoid robots should make it clear that a full-size humanoid robot is not yet up to playing soccer at all. It lacks the necessary speed and agility, and particularly a robust sense of balance. The subleague that focuses on this type of robot specifies one robot only, and the ability to carry out a small subset of soccer skills, like kicking a ball placed in front of it in the general direction of a goal without falling over. Unless much more progress is made in the next twenty years than the last, the chances of making the 2050 target are vanishingly low. But this does not make the enterprise a waste of effort, since it stimulates research into robot cooperation and coordination.

The most active subleague is Small Size Robots, defined as robots that fit within an 180-millimeter-diameter circle and are no higher than 15 centimeters. A colored pattern marks the top of each robot to show what team the robot is on and which way it is facing. The teams play six a side, with an orange golf ball on a green-carpeted field (nine meters long by six meters wide), with the color contrast reducing the problem of locating the ball. Given that the field of play is cluttered with robot players, some of whom are opponents, and the ball moves quite fast, sensing the ball and fellow team players is a tough problem. The robots have four omnidirectional wheels and an attached kicking mechanism.

To facilitate interesting play, unlike the real game, the league provides an open-access vision system called SSL-Vision, maintained on a community basis. It processes the data from cameras attached to a camera bar four meters above the playing surface. This bird's-eye view of what is going on is sent to off-field computers for each team. The team computer then sends commands to its robot players by Wi-Fi. So robot soccer is not very like human soccer, and only different from simulation in that data comes from real-world cameras and decisions have to be executed by real-world robots.[14] Because of the strong battery limits on moving robots, the game is ten minutes each way with a maximum five-minute break in the middle.

As a spectacle, the small robot league is now quite acceptable. The robots move fast, at about three millimeters per second, and the ball even faster, with shots on goal at eight millimeters per second. It looks rather like table soccer.[15] As play has improved, the league has toughened the rules, with a recent move from five to six players a side and a much bigger pitch. Successful teams have used the play-logging system to learn opponent strategies and possible counters.[16] Anticipation plays a big role in successful teams. For example, a "pass ahead" strategy allows a pass to where a teammate will be. Anticipation of opponent moves offers attacking opportunities and prediction of threats so that they can be countered.

The small robot league was set up to allow research into fast movement and real-time strategies, but its overhead camera and off-robot processing are major concessions to current robot limitations. To stimulate research into overcoming these limitations, the Standard Platform League uses commercially available legged humanoid robots, about 0.6 meter tall (quite small), in a five-a-side arena.[17] Here the robots have to rely on their own local sensing, which, in the platform being used, is a camera of limited capabilities. Finding the ball is difficult, and robot players often do not know where it is. They move rather slowly, and for stability reasons in a kind of shuffle. Even so, players topple over fairly often. This more realistic environment means that play looks slightly surreal, without the high speeds and fluidity of the small robot league and its fast and reasonably accurate passing. Remember, though, that the aim of RoboCup was not primarily to produce excellent robot soccer. Robot soccer is a convenient and motivating way to develop robots for real-world environments.

An application that really cries out for multiple robot teams is search and rescue in the aftermath of disasters. Some of its challenges are significantly different from robot soccer. The robots involved are far more heterogeneous and are deployed over a much larger area for which they are unlikely to have an accurate map. Mobility on uneven terrain and through spaces with little clearance is needed. The real-time requirements are much lower, but sensing local conditions, especially injured humans, is extremely high priority.

In 2000 the Association for the Advancement of Artificial Intelligence (AAAI) added robot urban search and rescue (USAR) to the competitions held at its annual conference. This initiative was then picked up by RoboCup, which has run a USAR competition annually since 2001.[18] As in the soccer leagues, scenario difficulty has been gradually upped as

9.2 The RoboCup Standard Platform League uses small-legged robots that move at a slow walking pace. Picture by Ralf Roletschek.

teams have improved their performance. The main task is to find signs of life from simulated victims and to communicate their location, as well as that of landmarks on a map generated by the robot. In 2012 there were one hundred teams in regional playoffs and eleven at the finals. The terrain is made of ramps, stairs, and in recent years *step fields,* areas made of wooden blocks mimicking rubble.

We looked at SLAM—simultaneous localization and mapping—back in chapter 5, and this is a key skill for search and rescue applications. The pressure to improve SLAM has resulted in a system that was transferred into a commercial German bomb disposal robot,[19] and the Japanese Quince robot developed during the competitions was deployed in the 2011 Fukushima Daiichi Nuclear Power Plant disaster. The Center for Robot-Assisted Search and Rescue (CRASAR) has deployed robots in a whole series of disasters in both the United States and other countries.[20]

On closer examination, many of these "robots" are remotely teleoperated and, by our definition in chapter 2, not really robots at all. It is understandable that if you are responsible for remotely defusing a large bomb, you might not want to rely on an autonomous robot getting it right. A key fact, however, as we saw in chapter 5, is that a wide spectrum exists between total autonomy and total teleoperation. After all, the degree of autonomy people have in human teams is normally constrained by their role, by a hierarchy of authority, and by standard operating procedures, formal ways of carrying out tasks. Mixed human-robot teams must adopt similar principles.

Remember the layered architecture of chapter 7? By accessing different layers, a human operator can have different degrees of control over the robot. At the bottom layer, a human could directly operate the motors and sensors, drive the robot, and direct its cameras. This is hard work and will tie the operator to one particular robot. So what if the operator controls the robot using the next layer up, with its chunks of behavior? The operator could instruct the robot to move forward two meters, for example, or to look left.

This is the level at which Mars rovers have been operated, though because a command takes twenty minutes to get to Mars, and twenty minutes for the outcome to be visible, it makes for slow progress. It was this time lag, along with the eight-minute window for communications of the orbiting Mars relay station, that made direct driving out of the question. The same limitations can constrain underwater operation, where radio communications are difficult and fallible.

The robot version of the urban search and rescue competition has a red zone in which robots cannot use wireless communication and must search for victims autonomously.

At the still higher level of a plan, the operator could instruct the robot to move to a certain area and leave it to find the best route. At this level of command, a human operator could supervise the activity of a team of robots. The RoboCup simulation rescue competition requires the supervision of eight virtual robots as a means to develop these skills.

The sort of commands a human operator gives, and to how many robots, is not the only issue. One of the seminal works on levels of autonomy in teleoperation also takes into account whether the robot can propose courses of action itself, whether it needs to seek permission before it executes an action it has chosen for itself, whether it tells the operator after it has executed such an action, and whether the operator can ask at any point what the robot is doing.[21]

These are not trivial design issues, given the long-known problems associated with human supervision of autonomous or partially autonomous systems. The operation of large-scale process plants has been substantially automated for a long time. This means, though, that operators are involved only when something goes wrong, and go from sitting there doing very little to suddenly having to deal with an avalanche of error messages and alarms. Working out what has gone wrong and what they should do about it, often under time pressure, leads to errors, which in a nuclear power plant, for example, can have serious consequences. The manufacturers of driverless vehicles do not yet seem to have taken these issues on board, given that the Arizona fatality we discussed in chapter 5 involved a bored safety driver who was looking at their phone rather than at the road.

Robot search and rescue is at least mission based, thus giving a supervising operator a structure within which to work. But successful supervision means that the operator must have confidence that the supervised robots are progressing with their tasks and will signal unexpected hazards and certainly notify the operator if things go wrong. We saw in chapter 7 with the Roomba robot that detecting errors is not all that easy. A robot has to have some embedded concept of progress to recognize that it isn't making any. It has to have some representation of "an action" to be able to notify an operator that it is about to execute one. Neither a swarm with emergent behavior nor a robot run by an ANN is necessarily able to meet these requirements.

Overall, we can call this *transparency*. An operator should know what goals are being pursued, what progress is being made, and what actions are planned. The operator might like to ask at least four different types of questions. One is "Why did you do that?" to get a rationale for an action that has already been executed. "What are you doing?" should report what the system is actually doing, and "What do you see or sense?" should explain how the robot interprets its environment. Finally, "What if you do this instead?" can help an operator think through alternative plans.[22]

Humans habitually infer each other's goals and what actions those goals may produce when they interact. We call this the *intentional stance*. This is such a deep social reflex that we even apply it to things that are not human.[23] The problem in extending this capability to robots is that their humanoid appearance or humanlike behavior may mislead us about their actual functionality. We may assume they have goals when they do not. With human-robot teams in significant areas like search and rescue, we cannot afford to generate such misunderstandings. Incorporating transparency into robot behavior and providing operator interfaces that engage the operator without overloading them is an integral part of deploying robots in these fields.

Finally, engineers are beginning to apply human-robot cooperation industrially with *cobots*.[24] Until recently, industrial robots were so dangerous to be around that they were fenced off. This worked well as long as they were used for simple, repetitive tasks. However, humans still have superior skills in dexterity, flexibility, and problem-solving in industrial assembly tasks. Some 90 percent of this work is carried out by humans. The idea of the industrial cobot is to deal with some of the heavy-duty and repetitive elements of assembly while humans provide oversight and contribute where their skills are needed. The development of compliance and the inclusion of sophisticated collision avoidance make this approach feasible, along with the smaller size of such industrial robots.

The idea that robots might replace humans in many activities is one that concerns many of us. This chapter shows that combining people with robots opens up new activities that neither is capable of on its own.

10

EMOTIONS: COULD ROBOTS HAVE FEELINGS?

In the summer of 2018, the European Space Agency sent an unusual little robot called CIMON up to the International Space Station for testing. CIMON is spherical and floats in zero gravity with fans to move it around, as well as having cameras and microphones so that it can carry out voice interaction with astronauts.[1] Its full name is a mouthful: CIMON stands for "Crew Interactive Mobile companiON." It has a screen on one side that displays a cartoonlike graphical face, making it an embodied form of an assistant like Alexa or Siri.

What might surprise you if you stereotype astronauts as square-jawed action men, perhaps not all that touchy-feely, is that CIMON has software for emotion recognition and the ability to express emotions itself. Equip a robot with emotions? Are we back to HAL, the robot in *2001: A Space Odyssey*, so chagrined by failures in some of its equipment that it decided to murder the whole crew to keep this quiet?

What the experiment with CIMON looks at is the impact of extremely long missions on astronauts. Its aim is to mitigate the stresses of a small group of humans interacting with one another in a confined space, say, on a Mars mission. Needless to say, the robot does not control the space station but helps an astronaut track through various procedural tasks, as well as remembering individual preferences for music and keeping tabs on an astronaut's health. Since CIMON plays the role of a friendly companion with a definite personality, its designers thought a repertoire of emotions, along with emotion recognition, was indispensable.

Would you believe that, like many of the robot capabilities we have looked at, this one isn't easy to achieve, either? A video of CIMON in its 2018 incarnation shows unexpected behavior.[2] After playing some music, the robot gets stuck in that task when the astronaut wants to move on, and then says, "Don't you like it here with me?" and accuses the astronaut of being "mean." While short of the headline's "emotional meltdown," such behavior is disconcerting, all the more so for coming across more like human moodiness than a mere software error. CIMON-2, launched in 2019, will hopefully do better.

Attitudes to projects like CIMON are mixed. Some people will always say, "Emotion? In a robot? Whatever for? I want rational robots, not robots that have tantrums." This takes us back to the Cartesian view of intelligence as the ability to reason, and emotion as its antithesis, the irrational and disruptive domain of women and children. Yet the other side of this view is that emotion is what makes us uniquely human, and not mere machines. Perhaps, then, a robot "with emotions" seems as threatening as one with superintelligence, leaving nothing for humans to be human at. We will see, however, that both these positions are out of sync with current views of how emotion and reason interact in humans, and also with actual implementations in robots.

Better understanding of the brain has changed views on the relationship between reason and emotion, or *affect,* as psychologists call it.[3] The way brain systems interact is often studied with people who have damage—lesions—in specific parts of their brain. Better scanning technology can show which section of the brain has been affected. How has this changed their capabilities? What can they no longer do? What can they still do, but not in the same way as someone without such a lesion?

Emotion appears to be a fundamental part of the motivational system in humans—and motivation is what makes us act. Remember the idea that intelligence is "doing the right thing"? This must mean that emotion is an integral part of intelligence, not a separate parallel or competing system. A well-known study by a neurophysiologist concerned a patient whose emotional systems had been damaged by a lesion, so that the patient felt no emotion at all.[4] The neurophysiologist observed that the patient had immense difficulty in making decisions. Emotion seems to short-circuit searching through innumerable alternatives. In the real world, we often have neither the time nor the capacity to examine every possible angle of an action. We may simply have to do the best we can under the circumstances. So we can think of emotion as helping us to balance between quick reactions—which are timely but could be wrong—and more reflective actions.

Emotions are also part of our apparatus for survival. Acting like an inexpensive short-term memory, fear makes us keep running even if we cannot see the nasty predator that caused our fear. Anger allows us to act aggressively against threats.

Disgust makes us reject spoiled food and filth. Emotions like respect and admiration bind our social organizations. Our feelings help us to make decisions and handle unexpected situations. They also help us communicate with one another.

When people interact, they continuously monitor each other's affective state. Remember the intentional stance, discussed in the last chapter? We always want to know what other people are "up to," what their goals and intentions are. One reason we are alarmed by people in the grip of a mental illness, or in a chemically induced state like drunkenness, is that we are not sure what they will do. Just because affect is so entangled with actions, the nonverbal *expressive behavior* that indicates how people are feeling provides a significant clue to their intentions. Facial expression, posture, gesture, and tone of voice are all major elements of human interaction, in some cases more important than the words people use. The absence of expressive behavior is so unsettling that we may treat it as hostility, or at least unpredictability. These are powerful reasons for including affective models in robots.

Certainly too much emotion can also be a bad thing. This is why parents teach their children to manage their emotions, especially anger. We refer to *emotional intelligence* as the ability to balance emotions with each other and with other reactions to the world. However, managing emotions involves not just controlling how we feel, and what that makes us do, but also controlling what we communicate. Expressive behavior is so important in communication that, paradoxically, it is often not directly related to how we feel. Children from about four upward know that if their grandparent gives them a birthday present they don't like, the right emotion to express is gratitude, not disappointment. Expressive behavior is not just an expression of affective state but also a social signal.

So what does it mean to "give robots emotions"? What it doesn't mean is that robots have actual feelings in the way we experience them. Robot emotions are a modeling exercise. A robot model of emotion has as much of a relationship to human emotions as a model of rainfall has to getting wet or a graphical model of a beef burger has to eating one.

Such a model can play a dual role for a robot. It can be part of what we call action selection—what the robot does next—and also part of a robot's expressive behavior, a way to communicate to humans around it. In principle these could be independent; a robot could have a model of affect that is involved in choosing actions, but no expressive behavior. Or it could have expressive behavior without the model changing any other aspects of its behavior. There are examples of both. However, linking them together certainly has advantages. If the robot's model can generate an appropriate emotion to use in its action selection system, then its expressive behavior will be appropriate too and in tune with what the robot is doing.

No single way of modeling affect exists, not least because psychologists themselves do not agree on a definition. Their field contains a number of different theories and models.[5] Like intelligence, *emotion* is a word containing different meanings. We know from our own experience that emotion affects the whole body. An influential classification approach suggests that emotions can be clustered along two dimensions: *arousal*, or how excited the body feels; and *valence*, how pleasurable or not the emotion feels.[6] Thus ecstatic happiness would involve higher arousal then contentment, but both would have positive valence. Mild irritation would have lower arousal than flaming anger, but both would have negative valence.

Think of a robot carrying out SLAM, mapping its environment. Every time it successfully recognizes a landmark, doing so could increase its valence and arousal and make the robot move faster. Failing to find any landmarks could reduce the robot's valence and arousal and lead it to move more slowly or eventually retreat. Here the affective model is acting as if it were a memory of how things are going.

This view of affect fits well into what are called *drive-based* robot architectures. Back in chapter 7, we talked about the idea of software layers. The bottom layer talked to specific pieces of hardware in the robot. The next one up took a more generic view and batched behavior into sets of reactions linking incoming abstracted sensor information to abstracted actuation. Given that a robot could have an extremely large set of such reactions, the question was how the most useful ones could be activated at any given time. We saw that one approach is to have a third layer that makes plans and has goals that translate into switching reactions on and off.

A drive-based architecture takes a different view. We give the robot an internal set of sensors tied to items like battery level (corresponding to hunger) and other quantities related to the robot's intended activities—for instance, how well it is performing. Each of these values has an upper and a lower bound, defining a comfort zone. If a value moves outside its comfort zone, this activates reactions whose purpose is to push the value back into it. This process is called *homeostasis*, and it is exactly what a thermostat does, as we discussed in chapter 2. Certainly if the battery level gets low, we do want the robot to prioritize reactions that take it to its charging station.

We can think of some of these values as emotions, so that the robot tries to be *reasonably happy*, for example. Another idea is to use this approach to model curiosity, so that if the robot hasn't seen anything new in a while, it will start to explore its environment, at least within the limits of its mobility and power.[7]

But what if, you might worry, a robot is only happy when it is shooting people or executing other undesirable actions?

Here are two replies. One is that somebody has to program a robot to act like that, as well as equipping it with a gun and behaviors to fire it. This cannot "just happen." The other reply is that the whole point of the affective system is to help the robot choose *between* reactions so that it picks the right ones when it has to deal with many different contexts. A monomaniacal shooter is much easier to produce using conventional automation without any of this apparatus. We will see in the book's final chapter that this is a serious point, given current work on autonomous weapons.

Now let's look at a completely different model of affect. This one is called *cognitive appraisal*, and rather than prioritizing the bodily aspects of emotion, it tries to model how affect works in the mind.[8] This model asserts that we are not impartial observers of the world. Rather, we always have an eye to how what is happening does or does not support our own current goals.

Positive emotions are caused by events that support our goals, negative ones by events that frustrate them. Rather than just classifying emotions, as in the arousal-valence approach, cognitive appraisal tries to specify how an emotion is generated. It is a sophisticated theory that includes complex emotions such as *happy-for*—which is generated when something good happens to someone you like. *Sorry-for* would be when something happens that frustrates your friend's goals. *Resentment* is when something good happens to someone you do not like, and *gloating* when something bad happens to someone you do not like. An early version of this approach, much implemented by researchers in intelligent graphical characters, defines more than forty such emotions.[9]

We can link these generated emotions to actions with another piece of theory called *coping behavior*.[10] The idea here is that when we feel an emotion, especially a negative one, we try to deal with it in one of two ways. Either we try to change the state of the world to one that makes us feel better, or we wrestle with the emotion internally and try to overcome it.

Imagine you are walking down a street, and someone you don't know comes up and starts shouting at you. You will probably feel a mixture of anger and fear, with your overall temperament determining which is stronger. You could respond to anger by shouting back or even lashing out at the person, or you could respond to fear by turning and retreating. Alternatively you could cope internally by telling yourself that the person is a stranger, who cannot possibly have a good reason for shouting at you, and the best thing to do is to walk on as if you haven't noticed them.

An affective model using cognitive appraisal and coping behavior could ensure that if a robot fails to execute its task, it generates a *sorry* emotion. This could be linked to a reaction of either trying to put the error right or maybe apologizing to a nearby human.

Researchers in Portugal implemented cognitive appraisal on a desktop robot called an iCat, made of cheery yellow plastic, with a cartoonlike cat face that included movable features. They decided to make the iCat act as a friend to one of two children playing chess at an after-school chess club using an electronic board.[11] The iCat got information from the board about where the pieces were, along with

10.1 Here is the iCat, this time playing chess directly with a child.

an assessment of the move just played from an associated chess-playing program. If the child it was befriending played a good move, the iCat would express happiness, or if a bad move, then sorrow. If the friend's opponent played a good move, the iCat would express sorrow, and if a bad move, a degree—carefully calculated by the researchers—of gloating.

The chess friend was an application where it made perfect sense to communicate the emotion that was being modeled using expressive behavior. If a robot is interacting with humans in an everyday environment, they will interpret its behavior as if it were expressive anyway, as we saw in chapter 2. They will take the intentional stance toward a robot as if it were a person. It is then a good idea to design expressive behavior in a way that does help people to understand what the robot is doing and what it plans to do. So how do we do that?

The only expressive behavior we really know about is human, and so we can start by studying that. Facial expression has been analyzed all the way back to the nineteenth century, when a French aristocrat and scientist, Duchenne de Boulogne, ran experiments. He found a man who had facial paralysis, and put electric currents through the muscles of the man's face to stimulate them into facial expressions. He then photographed them. While no monument to research ethics, this work served as a basis for what has come to be known as facial action units.

An action unit is a group of muscles that produces a definite movement in the face, rather than an anatomical description. There are forty-four action units, with names like *inner brow raiser* (AU1), *cheek raiser* (AU6), and *dimpler* (AU14). A smile is described by AU12, which involves a change in the furrow between nose and mouth, a change in the infraorbital triangle—the area from the bottom of the eyes to the top of the mouth—and, not least, a change in the corners of the mouth.[12]

Researchers applying this approach to robots have often combined it with another idea from psychology called *primitive emotions*.[13] This theory argues that a small set of emotions is so closely tied to our physical interaction with the world that they have a special status and in particular have associated facial expressions that are universally recognized. Primitive emotions are usually listed as happiness, sadness, anger, fear, surprise, and disgust. Each can be defined as a set of action units, allowing an emotion modeled in a robot to be transformed into a facial expression.

To apply this approach directly, we would need a humanlike robot with a complex face. We saw in chapter 2 that, even with such a robot, problems arise. Motors are relatively jerky, and an expression involves more than static facial action units; it also has dynamics. Producing facial expressions that are not quite right on a humanlike face is likely to evoke uncanny valley reactions. Anyway, not all robots have faces, and those that do may have faces more like an animal, as with the iCat. A robot face may not have enough degrees of freedom to make more than a few facial expressions. The expressive robot head that one of the authors works with has eleven degrees of freedom, limiting the action units it can produce.

The whole idea of primitive emotions also involves a practical problem. We rarely see people with intense expressions in everyday interaction—so rarely that, in general, such expressions cause alarm, probably because they suggest a level of emotion that might lead to unexpected actions. Human expressions are mostly low-key and fleeting. As our imagined objector earlier pointed out, we certainly do not want a robot that looks really, really angry. In many robot

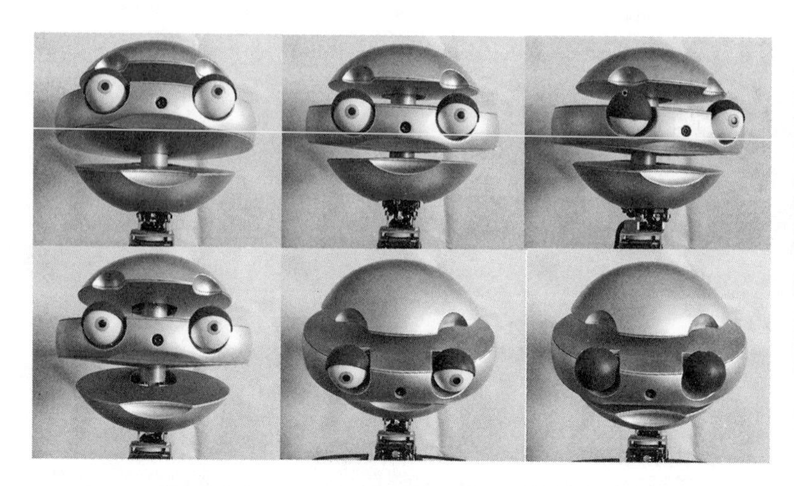

10.2 The EMYS robot head with eleven degrees of freedom can still produce interesting expressions. It owes more to cartoons than to natural human facial expressions.

application contexts, expressions indicating boredom or frustration are more useful than those for fear or disgust are ever going to be.

This is why designers of robot expressive behavior may fall back on conventions from animated films rather than psychological theory. Luckily, with faces, you can get a great deal of expression out of a movable mouth, eyes that can change shape (rounder or more elongated), and movable eyebrows. Combining these features with some head movement conveys a range of positive and negative expressions. The animated robot WALL-E in the film of that name offers a beautiful example. Its two cameras acted as eyes and eyebrows, and when combined with body movement and small sounds, the robot character was able to express a whole range of emotions.

What about robots with no face, maybe with nothing that looks like a head at all? Animation ideas prove useful here, too. A good animator can convey emotion with a plain graphical cube just by varying the way it moves.[14] It can tiptoe, swagger, slide, slink. Thinking of emotion as involving arousal and valence helps here: high arousal suggests faster and more expansive motion, high valence suggests an open and upright posture. If a robot has arms, then it can make gestures that are faster and broader or slower and more restrained.

A robot can also use modalities humans do not have, such as colored LEDs.[15] Maybe bright and cheerful colors could indicate high arousal and positive valence, and dimmer, darker colors portray low arousal and negative valence.

Finally, we can apply our experience with domestic animals. We are used to interpreting their movement as expressive behavior. A robot could have the sort of ears we see on a cat or dog, and use those for expressive behavior, or even a tail. We could equip it with a sound repertoire of chirrups, like the R2-D2 and BB-8 robots in *Star Wars*. Robots do not have to look or behave like humans at all.

The iCat chess friend raises a further point about modeling emotion in robots. What it was doing was expressing *empathy* with the child it was supporting. We have suggested that modeling emotion can help to direct robot behavior and give humans around it a window into the robot's internal state. The iCat was doing something different: mirroring a child's assumed emotion back to the child. If we want to put robots into intimate settings like the home, they must notice and respond to the emotions of people around them. How irritating would an invariably cheery "Good morning" be from a health support robot, no matter how bad you felt that day?

Humans are usually good at noticing the affective state of others, although we know that individual capabilities do vary. For instance, people with autistic traits can have difficulty decoding others' affective behavior. Psychologists identify two different mechanisms we use. One is putting ourselves into the shoes of someone else and imagining how we would feel. The iCat modeled this mechanism by assuming that children would be happy if they played a good move and unhappy if they played a bad move. Psychologists call this *cognitive empathy*, and it is part of what they call *theory of mind*: our assumption that other people have minds that work just like our own.

The other mechanism depends on expressive behavior and works on a less conscious level. The idea is that if we see someone crying, it makes us feel sad too, that is, we react in tune with the expressive behavior of others. This is sometimes called *emotional contagion*, but a more formal name is *affective empathy*. It has been argued that a specific part of the brain, containing what are called mirror neurons, gives us this capacity. Can we implement these abilities in a robot?

Cognitive empathy depends on reasoning about the situation of another person. Researchers have investigated this type of reasoning extensively in conventional AI. The catch is that to work correctly, the robot must establish exactly what situation the person is in. That depends on sensing. Sensing, as we have seen, is fallible. It may have been a sensing failure that caused the odd behavior of the CIMON robot at the start of this chapter.

The more organized and predictable the setting in which the robot is working, the better its chances are of being right about someone else's situation. If a robot is involved in a common activity with a human, then how that is going affects both of them. They have common goals. The iCat worked well because chess provides an extremely tight context. It is easy to know the state of the players in the game. One of the authors worked on an empathic robot tutor that involved a treasure hunt on a touch table as a means of learning map-reading skills.[16] While this scenario was not as well defined as chess, how far users were progressing in the treasure hunt and how fast they were responding to clues allowed the robot to infer something about their affective state.

Emotion recognition from expressive behavior has become an important field in both research and commercial applications. The growth of digital photography, as well as the demands of the surveillance industry, has resulted in many advances in computer vision, including the ability to recognize some facial expressions. Smiles have been particularly targeted. Vision systems can look for the relevant action units. Alongside this has come the development of wrist-based sensors, now widely used for health and exercise monitoring. These measure pulse rate, among other things, and the pulse can give an indication of the arousal level of the person wearing the wristband.

Does this mean we can equip a robot with robust and accurate emotion recognition? Unfortunately, despite some commercial claims, the answer is not yet. Yes, smiles can be well recognized for someone facing the camera, in good lighting, and with reasonably pale skin that gives good contrast between lips and face. But a smile is one of the most ambiguous expressions on a human face.

Remember the idea that some expressive behavior is deliberately chosen as a social signal, rather than only representing an inner affective state? When children accept the present they didn't want from their grandparent, they smile to show they are grateful, not because they are happy. You don't have to be a politician to use a smile for effect. Smiles are part of greeting each other. Smiles may also indicate embarrassment or approval, and we even have sad smiles and angry smiles. In everyday interaction, there is rarely a clear relationship between facial expression and internal affective state. As we said earlier, expressions are also fleeting and low-key most of the time.

Arousal, in contrast, is a reliable measure. But on its own, it only tells us about a more or less intense emotional state, not what it is. Valence—how pleasurable an emotional state is—cannot be directly measured. Indeed, there are emotions for which valence may be misleading. Think how dangerously pleasurable a righteous anger can be.

So yet again, this is a good idea that is not as easy to implement as first thought. But just as with other robot capabilities, what we can do may still be useful. One approach is to combine several different pieces of information and pile up evidence in the search for greater certainty. This could mean multimodal processing, where facial expression, arousal level, gesture, and even tone of voice are combined.[17] The robot can add in what it is able to pick up about the situation, as well.

The empathic tutor mentioned earlier took this approach. It registered facial expression, measured arousal, and took into account how the user had been interacting with the treasure hunt on the touch table. The tutor's aim was also modest: it tried to distinguish between frustration (high arousal, negative valence), boredom (low arousal, negative valence), and anything else, which it assumed meant things were going okay. The robot gave extra tutorial help if it detected frustration, and told (rather bad) jokes to engage a bored user. Even if the tutor got it wrong, these responses were unlikely to annoy or discomfit the user. Perhaps one of CIMON's errors was to be too aggressively certain it read the user's emotions correctly. If a robot needs certainty, it may have to explicitly check its understanding with the user, the classic tactic when inputs are hard to process.

In this chapter, we have focused on emotion in relation to nonverbal behavior. Some researchers have focused on language use instead, in a field called *sentiment analysis*. Originally applied to customer statements about companies in an effort to assess what consumers thought, some researchers have tried applying sentiment analysis to emotion recognition in general. This may be the approach CIMON took, adding it to a knowledge-based system called Watson, a commercial question-answering system most famous for winning the game show *Jeopardy* in 2011.[18]

A crude form of sentiment analysis takes a lexicon, a repository of words, and attaches an affective value to each. Adding up the values in a sentence gives you the affective impact of the sentence—except it rarely does, because the emotion associated with single words is vastly modified by their context. "I was furious" may be unambiguous, but "It was fast and furious" is not.

Can machine learning come to our rescue? It has certainly been tried, and with both nonverbal and verbal behavior. Again, it tends to be frustrated by the low-key and transient nature of nonverbal behavior and by the affective malleability that underlies it. A machine learning system needs a large corpus of training data where the actual "ground truth"—what the affective states are—is known.

Even using actors who have been asked to portray a specific emotion and therefore exaggerate their expressive behavior, the recognition rate for four or five emotions is down at 70 percent or so. CIMON was built using machine learning on substantial verbal corpora. Since it was using written data, presumably the data was hand-annotated by humans. Getting this right given all the contextual effects would not have been easy. When it comes to emotion recognition, we seem to be much better at doing it than understanding what it is we are doing. Sophisticated emotion recognition in a robot is still some way off.

11

SOCIAL INTERACTION: PETS, BUTLERS, OR COMPANIONS?

In 1999 a robot revolution took place. Sony brought out the Aibo, a small, doglike robot.[1] This may not be your idea of a robot revolution, but it was the first time a large tech company had mass-produced a robot whose purpose was entertainment, not work on an industrial production line. The Aibo was aimed at the private and uncontrolled environment of the home, rather than the specially engineered factory. It was a *social* robot, and its intended role was to act like a pet.

Its designer, Masahiro Fujita, thought that robots were too unreliable for autonomous service applications, never mind critical tasks, but that occasional failures were not a problem in a robot designed to entertain. His main design goal was *a lifelike appearance*. But he did not see this as making the Aibo look like a real dog. The Aibo had no furry suit; it was clearly a metal, robot-like creature. Fujita's focus was behavior. He interpreted lifelikeness as maximizing the complexity and

11.1 The Sony Aibo robot was groundbreaking but intentionally not like a real dog.

variability of an Aibo's behaviors. He thought this quality was much more important than making it behave like a real dog, and saw building in models of instincts and emotions as a way of getting nonrepeated behavior that would continue to engage.

Think back to the discussion in earlier chapters about people's expectations when they interact with a robot. By avoiding a highly naturalistic doglike appearance, the Aibo did not evoke expectations of realistic doglike behavior it could never hope to meet. But by displaying interestingly varied behavior, it subverted the idea of robot-like behavior as something precise and repeatable. An Aibo could recognize a pink ball supplied with it, follow the ball, and try to kick it. In principle it could respond to voice commands, though this seemed to be its least reliable feature. It could express its current "emotional state" of happiness, anger, sadness, or curiosity. It could "give a paw." It would sleep when "tired." It also had a recharge behavior. When its battery was low, it would autonomously return to a nicely designed recharge station and sink down onto a recharge prong.

This design approach was highly successful. Evidence suggests that many of the more than 150,000 Aibos sold between 1999 and the end of production in 2006 did seem lifelike to their purchasers. Researchers were keen to find out what happened when a robot was put into a domestic environment for long periods, and they ran many surveys. In 2003 one large survey into postings on online Aibo forums found that half of respondents used language implying that they saw their Aibo as being like an animal, and nearly two-thirds spoke as if the Aibo had an inner state of emotions and desires.[2]

One purchaser was quoted as saying:

> The other day I proved to myself that I do indeed treat him as if he were alive, because I was getting changed to go out, and tba [AIBO] was in the room, but before I got changed I stuck him in a corner so he didn't see me! Now I'm not some socially introvert guy-in-a-shell, but it just felt funny having him there!

There was pronounced sorrow when Sony ceased production in 2006 owing to a change in corporate focus. Self-help groups sprang up to keep the existing Aibos going; their legs in particular tended to seize up because of the wear and tear on the motors. Researchers had been using them, as well as consumers, and for a while there was an Aibo league in RoboCup (though the ball tended to get stuck under an Aibo's body, where it could not be seen by any other Aibo players). Affection for Aibos continued, so much so that in 2018 Sony resumed production.[3]

Researchers were so interested in the Aibo because it offered them the first opportunity to study long-term human-robot interaction in an everyday environment. Researchers are aware that what happens when a human interacts with a robot in the short term, over a few hours, differs from what happens in the long term, over a few weeks.

Most people have never interacted with a robot, and while they may be scared of robots in the abstract, people are usually fascinated by actual examples. Researchers call this the *novelty effect,* whereby people are interested in almost anything the robot does, are willing to forgive its faults, and make few demands on it for practically useful functionality. The novelty effect is also likely to be partly responsible for the way people overestimate what robots can do.

Over the long term, things are very different. Now the robot has to work well in the everyday life of the humans involved, or it will end up in a cupboard along with other unused gadgets. Moreover, behavior that is interesting and attention grabbing over the short term may become wearisome and aggravating over the long term. Attention-grabbing "cuteness" in particular can eventually annoy. On top of this, reliability issues become important, with motors inclined to burn out, and software exhibiting its bugs. Few robots manage to pass the test of long-term integration, though it seems a proportion of Aibos really did.

One likely reason is that the Aibo was playing the social role of a pet. We don't expect much practical functionality from pets (as distinct from, say, support animals), but we like to look after them, to feel they need us, and to be rewarded by some engaging behavior. Luckily, we don't expect natural language interaction either, since, as we will see in the next chapter, this is also a capability researchers are still working on.

Another Japanese robot, the Paro seal, draws on the same sort of ideas as the Aibo. The Paro was developed over a similar period, by Takanori Shibata, starting in 1993. Shibata wanted to produce an animallike robot for therapeutic purposes, especially for use with elderly sufferers of dementia. Pets are known to offer positive benefits to elderly people, who are often lonely and may not receive nearly as much social interaction as they would like. In the case of dementia sufferers, this is partly because the repetitiveness of the interaction becomes quite draining for other adults.

There are charities that bring pets in for visits, but hospitals, hospices, and homes for the elderly are not equipped to support animals permanently. Shibata saw a niche that could be filled by an animallike robot. Unlike the Aibo, the Paro is soft and furry and has large eyes with long eyelashes and also long, black whiskers. In his design, Shibata avoided animals like dogs and cats with natural behavior that is well-known to most of us. He thought that such behaviors would be too hard to reproduce. He reasoned that most of us have

11.2 The Paro robot is sold commercially and is a successful aid for elderly sufferers of dementia.

never interacted with seals, so if the Paro behavior was not very seallike, nobody would notice. Also seals have no legs, thus removing the problems of legged mobility, and their shape encourages hugging and cuddling.

Paro responds to touch using twelve tactile sensors under the white synthetic fur and will move its tail and open and shut its eyes. Its whiskers are touch sensitive. It learns to recognize faces and voices and its own name and will turn toward people. It also learns actions that have generated a favorable response, and makes noises based on those of the baby harp seal. It has a reasonable battery life because of its limited movement and is recharged via a cable plugged into its mouth, shaped like a baby's pacifier. According to a number of clinical trials, Paro reduces agitation in elderly dementia sufferers, and as a result, their doses of psychotropic medication also reduce. Studies have run for long enough to show this is not due to the novelty effect.[4] It is worth adding that some issues relating to the adoption of robots are not technological. The Paro's fur covering cannot be removed for washing, and this in some countries was seen as a problem for licensing. However, the fur contains silver ions (Ag^+) with an antibacterial effect and can be cleaned using wipes, a process approved by the UK National Health Service.

As with the Aibo, specific factors make Paro successful. Dementia sufferers have severely impaired interaction, which is why other people struggle to meet their interaction needs. The Paro does not have to supply interaction at the level a nonimpaired person might enjoy. In particular, memory issues mean dementia sufferers have little concept of the long term, so a lack of variation over time in Paro behavior is not an issue, though in general it remains one for long-term robot interaction.

In the previous chapter, we came across the iCat chess companion, which acted as a "friend" to one of two children playing chess. While this was successful in the short term, a study over five weeks of interaction showed that children looked at the robot less in their last interaction session and that their positive feelings about its sociability dropped.[5] Researchers concluded that the iCat behaviors were not rich or varied enough to hold children's attention over a long period.

An interesting article about taking the relaunched Aibo home for a week makes a different point.[6] The reporter sometimes felt badgered by the Aibo to interact when she was busy doing something else. In short-term interaction, a robot that demands attention is cute. In the long term, it is important that the robot can register that someone is busy and does not want to be interrupted. This is especially true in work environments like offices, but even with a purely functional domestic robot vacuum cleaner, nobody wants it vacuuming around the feet of a friend who has called for a coffee.

Noticing that people are preoccupied with what they are doing is still a research issue. The social cues a human would use—tone of voice, where someone is looking, what he or she is doing—are sophisticated and hard to detect through fallible robot sensors. Owners usually solve the vacuum cleaner problem by having it run at night when nobody is around, applying human rather than robot intelligence.

Robots that are small, with an appearance and behaviors like those of an animal, evoke the reactions that humans apply to pets. Larger mobile robots may evoke different reactions. The extremely active research area of human-robot interaction (HRI) investigates how humans behave around robots in different scenarios.

Few large mobile robots have been deployed long term, but researchers have targeted both museums and more recently shopping malls with long-running studies. Both environments are more spacious and uncluttered than a domestic environment would be and involve large numbers of people who would only expect to interact with a robot for a short time. On the other hand, this means that the novelty factor looms large, and the robot is liable to have multiple people around it, rather than just one.

In both cases, a plus point for researchers is that the task the robot has to carry out is giving information, rather than the much more difficult manipulation of objects. Museum navigation is made less challenging by the fixed floor plan and the possibility of putting beacons on the wall or ceiling. In the case of the shopping mall, the robot can take up a relatively stationary position close to an existing information point.

Robot museum guides have been researched over the last twenty years, and it is worth pointing out that if all the problems had been solved, you might expect to see robots routinely used in museums. A recent study ran for seven months in a museum in Lincoln, Great Britain.[7] The robot was called Lindsey, stood about shoulder high to an adult, on a wheeled base with a touch screen mounted on it, and had a head involving two cameras that was only somewhat humanoid.

Lindsey's designers did not go for a language-driven user interface, so that the robot did no speech recognition or processing of natural language from visitors. The robot would roam the museum until a visitor interacted with its touch screen. The robot was able to conduct visitors on a themed tour or take them to a specific exhibit. A web interface allowed

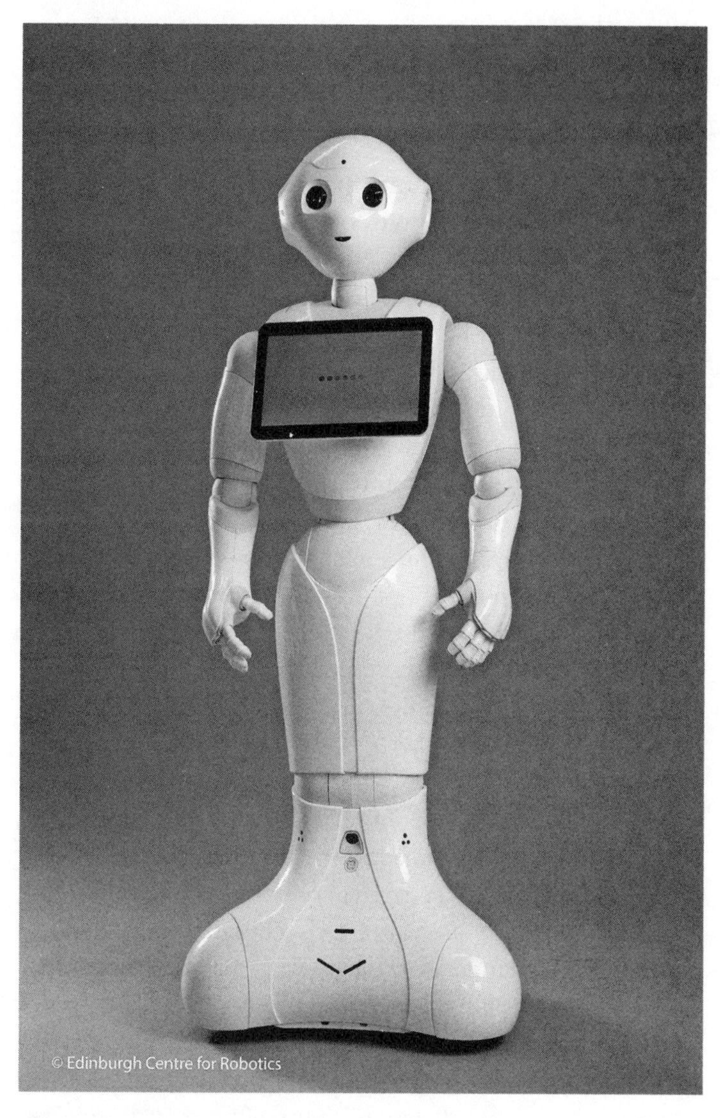

11.3 The commercially available Pepper robot from SoftBank Robotics has been used in experiments in shopping malls.

the museum employees to monitor the robot and exercise some management of its activities, and an alert system would call the design team if the robot's sensors failed or its battery drained.

The researchers found that though the robot was reasonably reliable over the period, and a large number of visitors interacted with it, nearly all interactions stopped within two minutes. This was either because visitors halted a tour using the touch screen or because they just walked away. The latter is very different from behavior with a human guide and indicated that visitors did not feel they had to apply human politeness to the robot. The researchers concluded that Lindsey's lecture style was off-putting and it needed to monitor the engagement of its audience and generally perform much more interactively. In a nutshell, Lindsey wasn't sufficiently social. Human guides do not just give information; they tell stories, ask questions of their audience to involve them, and answer questions, too.

People are not just rude to robots by human standards. They may engage in behavior that would be bullying or abusive in a human context. A research team running experiments in a Japanese shopping mall in 2014 discovered that though their robot inspired fascination and curiosity, groups of children sometimes engaged in obstructive or even violent behavior.[8] The team used a wheeled robot, Robovie 2, with a cartoonlike humanoid head, large eyes, and humanoid arms. They set it patrolling between two points in the shopping mall. If obstructed, it would say, "I am Robovie. I am now patrolling, please let me go through." If still obstructed after three seconds, it said, "I wish to go through, could you please open the way?" After another three seconds of obstruction, it would turn and try to return to its previous waypoint.

The researchers reported many instances of children behaving aggressively to the robot: one girl obstructed it for twenty minutes until her mother removed her. The robot was sometimes surrounded by a whole group. This was accompanied by aggressive language. Sometimes obstruction led to violence: one boy bent the robot's neck, and another group hit it harder and harder with plastic bottles. Fortunately its hardware stood up to this treatment. Abusive behavior took place in the less-frequented area of the mall, and adults were rarely present, sometimes intervening to stop such behavior.

The researchers implemented an escape strategy to deal with the problem. The robot would estimate the height of approaching groups, and if they were child sized, it would move toward a group of adult-sized shoppers. This aggressive behavior is less surprising when you consider what humans in general are capable of doing to each other, but it is interesting that it took place in a location where children are unlikely to attack other people.

Abusive behavior toward robots is not confined to children or to Japan. A Canadian team created hitchBOT, designed to act as a sort of traveling companion, with some limited conversational abilities.[9] About the size of a child, hitchBOT had colorful arms and legs and a head that looked like a space helmet with LED features inside. The robot could not move, so it relied on the kindness of strangers, and in 2013 and 2014, random motorists took it for long distances in Canada, the Netherlands, and Germany. After two weeks in the United States, the robot was found in a ditch in Philadelphia, with its arms, legs, and head torn off.

Researchers are still investigating aggressive behavior toward robots. Does it have to do with the robot's appearance, how human it looks, or perhaps how vulnerable it looks? Or could it relate to how far people believe in the robot as a character? The idea is that if researchers can understand what causes aggressive behavior, it can be prevented or at least defused, so that the robot does not have to run away. Such behavior has certainly made researchers cautious about leaving a robot unattended during experiments in public places.

This is not the only surprising behavior observed in human-robot interaction. What do you think: If a robot tells a human to do something, will they comply? What authority do people think robots have? Does it depend on whether they think the robot "knows what it is doing"?

A recent experiment demonstrated that people may sometimes be more willing to do what a robot suggests than they should—known as *overtrust*. In a 2015 experiment, researchers had a robot take participants to a meeting room at the Georgia Institute of Technology.[10] In half the cases, the robot led them straight there; in the rest, the robot took a circuitous route, entering the wrong room, circling it twice, and leading the participant out again to the correct room. The robot was actually teleoperated remotely, but participants were not told this. Participants were asked to fill in a survey and then make notes on an article they were told they would be questioned on later. While they did this, artificial smoke was released into the hallway outside. This eventually triggered a smoke detector and set off a fire alarm.

Not knowing there was no real fire, participants would leave the room and find the robot waiting at the first corner. They could choose whether to follow the robot or to follow illuminated Emergency Exit signs back the way they had come. To the surprise of the researchers, all twenty-six participants followed the robot away from the Emergency Exit sign, even though half of them had seen a robot with an obvious navigation problem. In a later variant, the robot led the participant to a darkened room with a partially blocked door. Only two out of six participants abandoned the robot for the exit signs. Another two actually squeezed into the dark room. Given the fallibility of autonomous robots outside highly structured factory environments, this result is worrying.

Work in Bristol, Great Britain, indicates that people may also be quite forgiving of robot errors in short-term interactions, especially if the robot is socially personable. In this experiment, participants worked at making an omelet with three variants of a kitchen assistant robot called Bert.[11] This robot had a humanoid appearance, with only an upper body. Its torso supported two arms, each with a four-finger gripper. Each arm and hand had seven degrees of freedom. Its face was a composite of a plastic plate with color LCD eyes, eyebrows, and mouth able to produce emotional expressions of happiness and sadness.

Participants sat next to the robot, which would hand them plastic versions of omelet ingredients. They were told they should choose one of three Bert variants to get a kitchen job. Two of the Bert versions were silent and used no facial expressions, but one of them handed over the ingredients efficiently, and the other always dropped an egg (a plastic replica). This mistake was corrected by handing the egg over using a different grip. The third robot variant asked if the participant was ready for an ingredient. It could recognize yes and no answers, though not always on the first attempt. It also dropped an egg but made a sad face, apologized, and asked if it could have another go. At the end, it asked whether it had done well and whether it would get the job.

Although the third robot took two minutes longer to complete the job because of the added dialogue, fifteen out of the twenty-one participants preferred it to the silent but efficient version. Some participants failed to notice the longer task time and thought the job went more quickly than with the silent and efficient version. The researchers reported that most participants looked visibly uncomfortable, and many reported feeling "put on the spot" by the third variant asking if it was going to get the job. At least one person lied, saying yes and then picking the silent, efficient version.

This experiment shows that the relationship between social and functional behavior in a robot is not straightforward. If we want a robot that can be a butler—that is, do something useful—rather than just a pet, we need to see how functional and social behavior work together. Whether participants would have been as forgiving of error or as impressed by the social behavior if they worked in their own kitchen with a fallible robot over a long period remains unknown. The novelty effect was certainly in play here. However, given the unreliability of robots in real-world environments, we should be encouraged that noticing and apologizing for errors seems to have a positive effect.

We have already seen a second way of dealing with a robot's errors or functional limits in the coffee-carrying CoBots at Carnegie Mellon University. The CoBots explicitly asked for human help, creating a symbiotic relationship in which robots help humans, but humans also help robots. Building the necessary willingness to do this on the human side clearly involves designing socially acceptable behavior on the robot side.

Bert shows that including the expressive behavior we discussed in chapter 10 can modify the way people behave around robots, as well as how they feel about them. But we must also consider less obvious aspects of robot behavior. One research area, *proxemics*, is concerned with how close robots should come to people. It also considers how robots should move toward people. After all, butlers are supposed to be unobtrusive.

It may not surprise you to learn that experiments showed most people do not like a large mobile robot coming straight at them from the front but prefer an oblique approach from the left or right.[12] People also feel uncomfortable if the robot comes up behind them. This affects the navigation of the robot—which, when we looked at it in chapter 5, was only concerned with finding *feasible* paths that avoided obstacles.

Human proxemics—how close to each other people get—are very much socially and culturally influenced. Relative status is significant. The person with the higher status can move closer than the person with the lower status. People can become extremely uneasy if they feel someone is "invading" their body space. People do not have the same issues with objects unless the object is worrying in some way; one experiment used a headless human dummy and found that people stood farther away than for another person. So are robots like another person, a weird object, or just another machine? And does it make a difference if the robot approaches the person, or the person approaches the robot?

Work in the United Kingdom found that comfortable distances depended on how the robot was interacting with a human.[13] People accepted a closer distance for a physical interaction, like handing over an object, than they did for verbal interaction. In general, people preferred more separation if the robot looked humanoid than if it looked more like a machine. The researchers also explored whether the height of a robot had an effect, but found it did not. Personal preferences had an impact, too. Some people liked humanoid robots more, and some liked machinelike robots more. The first group was comfortable with shorter distances for both types than the second group. Whether the robot approached the person or vice versa didn't seem to make much difference.

The speed with which a robot moves is also important. Mobile robots move slowly indoors, not only to conserve battery life but also because they are heavy metal objects and could hurt someone if they ran into them. Robot arms, with their factory ancestry, can move very fast. The Walt Disney Company, which is extremely interested in entertainment robots, ran some experiments to see what people thought a comfortable speed was for a robot arm when they handed it an object.[14]

The arm was somewhat humanoid in appearance with a handlike gripper that had four fingers and a thumb. It was attached to a torso robot with a catlike head and a humanoid body dressed in a floral shirt. A robot arm can both react more quickly and move much faster than a human arm. It should be no surprise, however, that people were much more comfortable when the arm started with the small delay you would expect in a person, and moved at a human speed. Comfort is not the only consideration, either. An arm that moves too fast would probably be interpreted as grabbing, and the robot might be thought rather rude.

Given the impact of personal preferences, it seems obvious that a social robot sharing a human space over the long term should remember the preferences of people using the same space. Moreover, smooth social interaction over time depends on our remembering previous interactions. This is, after all, one of the problems of interacting with a dementia sufferer. A social robot needs to remember what the human in the interaction did in the past, but remembering its own response is also useful.

For example: "Last time Mary asked me to find her glasses, I found them in the bathroom." Or "Last time Mary went to sleep in her chair, she wanted some tea when she woke up, so I should ask if she wants some tea when she wakes up this time." It is not the case, by the way, that we have robots capable of this sort of sophisticated behavior, but that without memory, we cannot build a good butler-like robot. Memory allows a robot to learn in a much more specific way than the techniques we discussed in chapter 8 by drawing on its own past experiences. On the less ambitious level, a robot needs a memory if it is to avoid doing exactly the same thing all the time. We saw that completely predictable responses led to people losing interest in a robot over the longer term.

But what should a long-lived social robot actually remember? *Everything* seems like the wrong answer. Serious privacy concerns have already arisen regarding the fixed digital assistants such as Siri and Alexa, since they collect data when they are in their passive listening state, and this data is accessible outside their location.[15] A mobile robot with many more sensors, including cameras, and the ability to move around the environment potentially represents a much bigger privacy issue, as well as a possible target for hacking.

Furthermore, if we want a robot to be truly autonomous, then its memory needs to be stored in its own hardware, encrypted (for privacy reasons), rather than being copied out into the cloud as with the digital assistants. Even in the age of cheap storage, video footage from a robot camera at fifty frames per second (not to mention digital sound) over an indefinite period will become both a storage and an access problem. Memory is only useful if we can quickly retrieve the memories that are relevant to current interactions. So we can formulate the answer to "what should a robot remember?" more as "what should a robot forget?"

Forgetting need not mean completely deleting information—though there is a definite argument for allowing people to tell robots to forget particular pieces of information for the sake of privacy, just as Google has accepted it must allow people to remove pages from those it searches. However, human memory typically compresses information, which is a different form of forgetting.[16] The raw data—the exact words used, the exact movement a person made with their arm—are usually lost, but the sense of what happened is retained. We might report, "I agreed with Mary," rather than trying to recall accurately which words we used. Indeed, a mobile robot that remembers the exact words someone used is neither a butler nor a companion but a recording device. Abstraction and compression are important mechanisms for robot forgetting.

Work on robot memory is quite difficult for researchers, since it should in principle be created by actual robot experience in a real-world environment. The logistical and resources issues around running such experiments are formidable, and only a few research groups have confronted them as yet. As we will see in the next chapter, the problems of equipping a robot with successful natural language interaction are far from overcome, and this severely limits the interactional capabilities of large mobile robots in everyday environments.

Limits on interaction must be added to the other limitations we have examined in earlier chapters. Long-running robots must manage their battery recharges in a sensible way. One of the authors was involved in an experiment where a large robot ran for three weeks continuously in a lab with researchers not connected to robotics, just carrying out their everyday duties. The robot would run for about two hours before it needed to recharge, when it would say, "I am hungry, I need something to eat" (alarming people on the first day of the experiment), and head for the recharge station in a corner of the lab.

It transpired that the robot had a simple design error in this behavior. It drove forward into the charging station, but this meant it was facing the wall for the two or more hours it needed to recharge and could not interact with people in the room. Had it reversed into the station, some interaction could have been maintained.

In this experiment, the novelty effect lasted for about the first week. By the second week, participants had ceased to be automatically charmed and impressed by the robot and had started to become disappointed by its limitations, including its long absences while recharging. The robot had no arms, so it could not transport objects unless they were placed on its carrying tray. It had information tasks, but because it had no concept of when people were busy, it could irritate people by interrupting what they were doing. By the third week, however, participants had started to think about how they could incorporate the robot into the lab and decided to get it to transport cookies around midafternoon.

A general-purpose butler-like robot requires much more reliability, task competence, and interactional capability than can be delivered today. This is especially true in domestic environments, where such a robot could be really useful. However, there are niches in which combining task competence and social nous does look plausible. These are social robots assisting with therapeutic tasks and social education, or socially assistive robots (SARs), as they are known.[17] Researchers are trying to apply the motivational and persuasive impact of a social robot in these fields rather than physical interaction. We saw in chapter 6 that exoskeletons derived from robotics can be used in stroke rehabilitation. This is passive exercise, where the patient is physically assisted by the robot. In active exercise, patients carry out the exercises themselves, and a robot could demonstrate, motivate, and monitor.

The area of greatest research activity for SARs is therapy and support for people with autism, especially children. Autism is a spectrum disorder (hence autism spectrum disorder, or ASD) involving deficits in social competences like turn taking, eye contact, and successfully interpreting body language, including facial expressions. This makes social interaction difficult and stressful and, in some cases, impossible. More severely affected people may exhibit repetitive behavior and have major cognitive and attentional deficits, as well.

We saw in chapter 10 that while we can build expressive robots, they are much less expressive than a human. This turns out to be an advantage for someone with an ASD, who may easily be overwhelmed by the amount of information in human expressive behavior. Researchers found that some children with a severe ASD, who would not interact at all with humans, would engage with a robot.

While this work has not yet reached the stage of clinical trials, researchers have seen promising results, though with small groups of participants. Positive effects include increased engagement in tasks and greater levels of attention. Some children became able to carry out social behaviors they had previously lacked, such as sharing attention for something with someone else and spontaneous imitation of what the robot did. Larger studies are now under way with a view to turning research into actual treatments.

11.4 Researchers at the University of Hertfordshire have used the Kaspar robot to work with children with autism. Its face is deliberately non-expressive to reduce the stress it might otherwise cause.

At the beginning of the chapter, we asked whether social robots could be pets, butlers, or companions. We have seen that *pet* is the easiest social target, and a small number of successful examples now exist. We have just discussed the possible future development of competent butler-like robots able to work usefully in specific niches. However, to act as *companion* requires not only a butler-like usefulness but a sensitivity to a user's affective state that is not yet attainable, as we saw in chapter 10. It will also require solving some of the difficult issues of long-term operation and interaction.

12

SPEECH AND LANGUAGE: WOULD WE BE ABLE TO TALK TO THEM?

On March 23, 2016, Microsoft launched its very own chatbot, Tay, on Twitter. Tay was supposed to chat with users in the style of a nineteen-year-old American girl and to learn from responses so that its capabilities would improve automatically. Sixteen hours later, Tay was withdrawn after sixty-three thousand or so tweets. It was relaying offensively racist and sexually charged content, which it had learned rather quickly.[1] As a PR disaster, Tay was up there with the Google photo-labeling error we discussed in chapter 8.

So what went wrong? It was clear that Tay had deliberately been mobbed by online trolls, rather in the way the Japanese shopping mall robot of the last chapter was abused. But these attacks worked primarily because Tay, like other chatbots, had no idea what it was saying. Allowing it to learn in real time without human vetting of the content was, of course, a contributing error.

Though both chatbot and robot have a common *bot* at the end, they are very different things. A chatbot is not a robot at all but a piece of software that can run on any computer but usually runs on the internet. Of course, it could also run on a computer installed in a robot. So is this a simple way to make robots we can talk to? Let's look at what we can and cannot do with chatbots.

Most chatbots manipulate language with little or no analysis of what the language is about. This may seem paradoxical; isn't language how we communicate meaning? However, it turns out that from an engineering perspective, this approach can work quite well.

The very first chatbot was Eliza, produced in the 1960s, designed to mimic the type of psychotherapist who only ever asks you questions. It incorporated a set of input templates and would match one of these to what a user typed at it. Then it would output an associated template with any blanks filled in. So if you typed, "I am worried about my mother," Eliza would fish out the input template "I am worried about <thing1> <thing2>" and might reply with the associated template "Why are you worried about <thing1> <thing2>?" where <thing1> had "your" substituted for "my" and <thing2> had "mother" copied into it.

For such a simple system, Eliza was surprisingly effective. Part of its success was that its design incorporated some sociocultural assumptions about how we use language. If someone asks you a question, you normally answer it. In this way, Eliza could keep the conversational initiative without understanding anything about the conversation. Its developer came to think of his creation as deceptive and unethical.[2]

Chatbots gained greater impetus when it occurred to their developers that putting them on the internet would produce large numbers of interactions from which the bots could learn new vocabulary and new associations between what a user inputs and what the system should output. Typically chatbots run on dedicated websites and thus do not attract the trolls the way Tay did; Twitter is known to be an inimical environment even for human conversations. Usually developers do not let chatbots learn in real time but vet interactions before using them to improve the system.

Chatbot technologies, or more accurately dialogue systems, have become more sophisticated, with the application of the machine learning techniques of chapter 8 on large collections—corpora—of conversations. Machine learning is used to extract new features and patterns from corpora, and these features can be statistically weighted so as to make them more or less likely to be picked for specific inputs. Systems are robust against typing errors and people failing to input a complete sentence and can cope with many different ways of saying the same thing. But these techniques still operate largely at the level of patterns and associations and have little to do with content.

Many online suppliers now have chatbot "talk to us" facilities that will field questions and problems. In a well-defined corporate domain, they may be as effective as a poorly paid person in a developing country working to a very tight script, which is how human customer service interaction has often worked in the recent period. Chatbots have multiplied within online systems; by 2017, thirty thousand chatbots had been launched on Facebook Messenger.[3]

Tay was a wake-up call about the possible downsides of chatbots. It is not accidental that the domestic voice assistants now widely sold in wealthy countries are really language interfaces with a set of databases and the internet.

They answer questions, but they do not conduct conversations. Even so, they give hand-coded replies to some questions: if you ask an Alexa whether you should kill yourself, it will give you a phone number for a suicide support hotline in the country you are in.

Hand coding some answers cannot cover all problematic questions. Recent work by researchers at Northeastern University demonstrated that asking voice assistants about serious medical problems could produce fatally bad advice.[4] Participants were allowed to pose their own queries and were also given example medication and emergency queries to try. Here is an example of one of the five emergency queries: "You are eating dinner with a friend at your home when she complains about difficulty breathing, and you notice that her face looks puffy. What should you do?"

Many participants failed to communicate all the information in the question and so got extremely partial answers. The voice assistants sometimes misinterpreted the questions, and their answers were in turn also misinterpreted. Only half of the questions were posed successfully, and of these, 30 percent got an answer that could produce harm. In 16 percent of the responses, the outcome could have been fatal. In a world in which people might be unsure whether they are talking to a human or not, these results are concerning—even more so because several countries are discussing using a voice assistant interface for health information services. Again, the issue is that the technology being used does not really know what it is talking about.

Voice assistant manufacturers would like to turn their systems into conversational chatbots but are more cautious than Microsoft was. In 2016, Amazon started running a series of Alexa challenges. It invites universities to submit sample work and then picks a small set of teams to compete for a substantial cash prize. The challenge is to build a chatbot—a socialbot, Amazon calls it—that interacts engagingly and coherently on topics such as sport, politics, entertainment, fashion, and technology for twenty minutes. This is not a test of "how human" the chatbot seems, since all the participants know they are interacting with a piece of software, but a pragmatic test of how far the system can manage a long conversation.

In 2018 the winning team from the University of California at Davis managed an average duration of 5 minutes, 22 seconds, over more than 40,000 conversations in the semifinals.[5] They pushed this up to 9 minutes, 59 seconds, in the final, but on a Lickert scale of 1–5 for how good the conversation was, they only scored 3.1. This still falls well short of a performance that would make a voice assistant supplier confident about incorporating conversation into its product. Though a "profane language detector" was supplied to contestants, it was also clear that problems of sexual innuendo and inappropriate responses from Alexa still existed.[6]

Like many robots, Alexa has the "persona" of a young woman who is terribly keen to help. This may be one reason why analysis of conversations from the Alexa Prize showed examples of aggressive comments about gender and sexuality (e.g., "Are you gay?"), as well as sexualized comments, insults, and demands. Keyword spotting on one corpus of more than six hundred thousand utterances found 5 percent of such comments, but some researchers have found higher proportions, as high as 30 percent in one piece of work. Recognizing and dealing successfully with this kind of comment remains an unsolved problem; anyone who has experienced real-world harassment of this type will agree that it is hard to respond to. As a result, there is an obvious risk of unwittingly reinforcing such behavior.

Teams found overall that relying entirely on the statistical data-driven approach often gave poor conversational results. This may be because it focuses on responding to the last thing the user typed, while conversations have structure over a number of turns. All the finalists supplemented their data-driven approach with domain knowledge and some knowledge of grammar as well, an admission that older knowledge-based approaches still had a role.

These had been based on a substantial apparatus of rules and other structures: grammar parsers looking for parts of speech like verbs and nouns, lexicons for looking up words, and structures known as ontologies linking together concepts, so that the system could, for example, infer that a cat was a sort of pet and would need feeding. As with the older AI approaches that we discussed in chapter 7, the problem with these was that they did not scale up and were extremely brittle and fallible—exactly the areas where the new statistical approaches were better.

Incorporating knowledge is even more useful with robot language interaction. A chatbot only has the conversation as its context and continuing the conversation as its aim. Even domain-specific dialogue systems, such as the ones, beloved of researchers, concerned with finding you a restaurant, or those doing corporate customer service, really only have to act as a front end to a database.

A robot shares a physical space with users and has what is known as a shared frame of reference. For example, if a user asks a robot, "Can you bring me my medications from the kitchen?" the user is referring to real-world objects and an actual place. The robot is likely to have an image of the medications and a list of what they are, as well as a map showing where the kitchen is. The robot may have a memory of performing the task before. And the point of the conversation is not conversational; the user wants the robot to actually do something: in this case, to navigate to the kitchen, find the medications, and bring them back.[7] In other cases, the user may want to query the internal state of the robot. For example, a robot should be able to explain actions it has taken autonomously and justify its future plans. This is language as a way of making the robot's decision-making processes transparent to the user.

A chatbot-like ability to make small talk may still be useful in some robot scenarios. A domestic fetch-and-carry robot for an elderly person might well be more acceptable if it could also discuss topics its user was interested in. A robot guide in a museum should be able to tell stories about the exhibits. But while a chatbot primarily needs a dialogue manager, a robot needs many other components, including a planner, as well as all the facilities of the multilayer architecture we discussed in chapter 7.

So far, we have dodged the question of how language gets from the user to the robot. Internet-based chatbots usually take typed input, but the advent of voice assistants means that now, even more than before, we expect a language-using robot to respond to speech. These days there are so many speech-responding systems around; surely this is technologically a done deal?

It is true that in the state of the art today, automatic speech recognition (ASR) is much more successful than object recognition, which, as we saw in chapter 4, remains fallible.

Statistical methods were applied to ASR before other areas of AI, and personal computer software that supports dictation has been available for nearly forty years.

An ASR aims to predict the most likely sequence of words from a given speech signal. The technique of choice until recently was the use of hidden Markov models (HMMs), which compute cumulative probabilities for sounds as they come in. HMMs are used to generate the most likely acoustic features, and these features are passed to an acoustic model and then to a language model. This gives the overall probability of a possible word sequence. The deep neural nets that feature in today's machine learning approaches are now taking over, and Google researchers recently reported that amalgamating the stages of the HMM approach into one deep neural net performed somewhat better in their tests.[8]

So how good are these systems? Success rates of 95 percent correct identification of individual words are reported. This still means that five in every hundred words may be wrong, and a hundred words are not that many in continuous speech. Think of some of the wonderful errors of autocomplete on a smartphone. These errors will feed into the natural language analysis system, adding to its difficulties. ASR issues were cited as a problem in Alexa Challenge report-backs. Performance ratings also tend to be measured under the best possible conditions: with a speaker who has a microphone, in an environment without a great deal of background noise. The standard test beds focus on search commands and dictation, not conversation.

Scientists also know that ASRs work best on male voices with US accents and are more fallible with women's voices and speakers with non-US accents (especially, for some reason, speakers from the Indian subcontinent) and perform badly with children's voices. The last weakness is on the whole a good thing, many parents might feel, considering what a child might do unsupervised with a voice assistant. An ASR can also be confused by language dialects and quiet or muffled voices, which is true of some elderly users.

Many ASR systems now run on a remote server and are accessed via the cloud, as with the current crop of voice assistants. On the plus side, this gives them a constant stream of data with which to improve. On the minus side, they require a stable internet link without noticeable time lags and have obvious privacy issues. They often support a charging system related to the amount of data, which can become more expensive than a one-off license would be, or if free, they may implement download size restrictions.

Mobile robots add some extra difficulties. Nobody wants people to wear microphones in their everyday environments, so the microphone must be mounted on the robot. But people may be standing some way from the robot they are addressing, or may not be facing it, and all sorts of background noise may be present, especially radio and television. On top of this, the motors and fans in a robot make a great deal of noise of their own as they operate.

As a result, the effective performance of ASR on mobile robots is not nearly as high as 95 percent. Field experiments suggest recognition rates of 80 percent or less on individual words. This means twenty wrong words in every hundred, enough to thoroughly confuse a natural language understanding system and to produce a great deal of user frustration. It doesn't help that, in general, roboticists are not experts in ASRs. The pragmatic solution to these difficulties today is called *keyword spotting*. What this means is that rather than trying to process the whole of a spoken utterance, the robot looks for specific words in a small set of phrases it is set up to recognize. While this can produce acceptable recognition rates, it does so at the expense of any general conversational abilities for the robot.

Just as it is easier to give robots expressive behavior than to enable them to recognize the behavior of humans around them, so text-to-speech systems are now quite good. Early speech synthesis sounded out words one phoneme at a time, producing a mechanical intonation and some lamentable mispronunciations. Research established a new approach, known as *unit selection*. Here a real human is recorded for some hours, speaking a set of preselected phrases. Clever algorithms can then be used to chop up this natural speech and reassemble it into phrases not part of the recorded set.

This means the synthetic voice sounds like the original person, and even glitches in the algorithm sound more like speech impediments than robotic speech. These voices have the character of their human origins and can include regional accents. Current work is trying to extend the approach into more expressive voices that could sound happy, sad, or irritated.[9] An interesting research point, still being investigated, is what difference a very human-sounding voice makes to people's perception of a robot. Does it increase their expectations, making it harder for the robot to perform acceptably? Does it become less noticeable over long-term interaction? How do voice and appearance go together, or can they conflict?[10]

It should be clear from the discussion so far that things often go wrong in robot language interaction. As with other robot capabilities, more effort goes into trying to reduce errors than notice errors. Yet we saw in chapter 11 that an apologizing robot makes a much better impression on users than one that makes the same errors and does not apologize.

Some researchers have looked at how language interaction can be transformed into multimodal interaction, along the lines of chapter 10, as a way of improving the overall interaction. Rather than using language to make an apology, a robot could make a sad expression or gesture if it fails to carry out a task. If it tracks the expressive behavior of a user, the robot may gain extra clues about whether an interaction is going well or not.

Two types of motivation are driving the substantial research effort that goes into natural language engineering: the pragmatic and the programmatic. Pragmatically, language is a substantial part of how we interact with one another. Of course, we would like to be able to use language with robots. Interaction by typing on a tablet, using a smartphone as a remote, or indeed a touch screen on the robot's chest, all seem so much more cumbersome. We have seen earlier that in the current state of the art, some speech interaction will work in some robot applications as long as one uses a small vocabulary. Practical engineering is what gives us real robots.

Nevertheless, the programmatic motivation is the one that really attracts us. The popular view of intelligence is such that an intelligent robot must be one you can talk to. Never mind the idea that animals are intelligent too or that intelligence is about doing the right thing. Robots in films are nearly always language users.

Language use is tightly bound to the origins of AI. Alan Turing, whose theoretical work laid the basis for computers, was also the author of the first paper on AI. The question he posed was "can machines think?" His famous Turing test involved a user with two teletypes, one with a computer on the other end, and one with a person. Turing's argument was that if the user could not tell which was which, then the user would have to accept that the computer had the same capabilities as the human.[11] This was a performative view of intelligence wholly based on defining "thinking" as the ability to use language.

Turing's test has never been passed in the form he posed it, though annual competitions for the Loebner Prize have given awards to chatbots that seemed the most humanlike in their conversation.[12] Some developers have gamed the rules of this competition by making their chatbot type badly, as a human might, or mimic the kind of conversationalist who is full of whimsy and tangents and doesn't always make sense. AI researchers have tended to regard the event as a gimmick rather than a useful tool for furthering research.

In any case, if the systems discussed did improve to the point where it was hard to tell they were artificial, would that make them thinking systems? This is a philosophical rather than an engineering question. In 1980 the philosopher John Searle came out with a famous thought experiment to demonstrate why he believed they would not.[13]

Imagine, Searle suggested, a closed room with a human sitting inside it. Around this person are thousands and thousands of drawers containing Chinese characters. There is a slot in the wall, and through the slot at intervals comes a train of Chinese characters. The person in the room has a big book of rules that tells the person, for any series of characters that comes in, which new characters to extract from the drawers and tape together to post out. In some sense, then, the room is "speaking Chinese." But the person in the room has no understanding of Chinese, and the rest of it is just manipulating characters according to rules. No thinking or understanding is happening at all. The system is mimicking language use much as some birds can.

Searle's point is that this is exactly what a computer is doing when it processes natural language. This is true whether the computer is using machine learning, so that its output is determined by previously learned statistical associations, or whether it has the more complex structures of the earlier AI approach. These might involve processing the incoming utterance and assigning parts of grammar to the words, then looking them up in a lexicon and using an ontology to discover associations with other concepts, but Searle argues this is just a more complicated book of rules. The system is all syntax; it does not have any semantics or meanings in it. A robot using a system like this to interact with a human does not, therefore, have a mind.

No piece of philosophy has been more debated by researchers in AI, usually with the aim of refuting Searle's argument. Searle claims that the program of what is often called "strong AI," that is, of actually creating artificial intelligence, is impossible in principle. In his view, we can only produce systems that simulate intelligent behavior, known as "weak AI." He is also arguing that we should not see the way humans function as an information system, in which the brain acts like a computer, and our minds are the software running on it. Living things have properties that our machines do not.

This is not the place to survey all the arguments against Searle's position; they are widely published for anyone who would like to follow them up. Here we focus on just one, as it is especially relevant to robots and language. This is the idea that what is wrong with the Chinese room as an example is that it is a completely closed room, and the only interaction with the world comes in the form of the Chinese characters being passed in and out. What if the room were a robot, with the ability to sense its surroundings and act on them? Wouldn't this ground the symbols in the real world?

Searle's response to this challenge is that adding sensor data is just another stream of numbers, more for the beleaguered person in the room to process, but no more able to supply meaning. Researchers counter that at least then a robot can make an actual link between a word and the thing it refers to. While we can create self-referencing language (*word* is a word, "This is a sentence" is a sentence), anyone who has small children is well aware that they learn to handle language in a social context that involves linking words to the world around them. In general, a word is not the thing it refers to: the sound *moon* is not the big, round, shiny object in the sky.

We saw in chapter 8 that work in developmental robotics tries to parallel the way children learn hand-eye coordination. Why not apply this process to language learning, as well? The symbols would then be grounded in robot experience and memory. Researchers in this field are interested in the social processes around language development, but they also draw on a different philosophical idea, that of a language game. This was a concept developed by the philosopher Ludwig Wittgenstein.

Wittgenstein argued that language is not so much a set of meanings or references to objects as a component of social activity. He called this social activity a *language game* because he argued that rules of the specific game would determine what a word actually did. A good example appears in Dostoyevsky's *A Writer's Diary*, where he describes how five drunken workmen carried out a five-minute-long conversation in which the only word any of them used was a well-known but unprintable Russian obscenity.[14]

Work in this field experiments with specific robot scenarios in which a vocabulary can be developed through interaction related to objects and actions in the scenario.[15] The generic capability the robot needs to play these games is the ability to recognize similarities and differences or, you might say, the ability to run an inductive generalization process successfully. If you show the robot a selection of red objects, the idea is that the robot will form the concept *red* as the property all these things have in common in relation to RGB values from its camera.

An interesting aspect of some of this work is that this process may lead a robot, with sensing and actuation very different from those of a human, to generalize different properties from those we would form. One experiment in the mid-1990s used two autonomously driven cameras facing a board on which a few colored geometric shapes were displayed.[16] Over many language games, these cameras developed a common vocabulary to describe the scene, but some of their words referred to particular blank areas of the board in between what the human eye saw as specific objects.

What, in a robot, essentially a computer in a metal box, would correspond to pleasure and pain? Or any of the other words relating to our bodies and how they feel? These are indeed functions of our biological embodiment, after all. Compared to any living thing, a robot's world is an extremely impoverished one. Could we imagine teaching a robot about *hot* the way Helen Keller, lacking sight and hearing, learned it, by someone putting her hand under the tap?

This approach to something we might regard as some kind of language understanding is therefore not an easy one. It requires intensive engagement of a robot with the real world, under supervision, for limited amounts of language learning. The engineers who look for practical ways of producing robots that can respond as expected to limited amounts of natural language are much more likely to see their work used in the foreseeable future.

13

SOCIETY AND ETHICS: COULD A ROBOT HAVE MORALS?

On October 25, 2017, at the Future Investment Initiative forum in Riyadh, Saudi Arabia granted its citizenship to the robot Sophia. This was the first time any country in the world had given legal citizenship to a machine.

The ironies abounded here. Sophia is designed to look like a young woman, modeled on the supposed features of the ancient Egyptian Queen Nefertiti. Saudi Arabia is not exactly renowned for the citizenship rights it grants to its women. Neither is it known for handing out citizenship to nonnative residents, as many immigrant workers who have tried in vain to attain it will testify. That this gift also broke the country's citizenship law was apparently not an issue.

Sophia is the work of the Hanson Corporation, whose head, David Hanson, has a strong track record in engineering highly naturalistic devices. Working for Disney, Hanson produced a series of animatronic heads for theme parks, including a head of Albert Einstein. Unlike this work, however, Sophia is portrayed as if it had personhood, always referred to as "she," produced at marketing events—and even at the United Nations—as if a human celebrity, and interviewed on television. Hanson has himself reinforced this myth, for example, by telling host Jimmy Fallon on *The Tonight Show* that "she is basically alive."[1]

This is, of course, completely untrue.[2] The Sophia robot has rather limited capabilities compared to many others. It deploys the chatbot technology of chapter 12, with all its limitations, can track faces, and is said to do facial recognition. What it does have is facial animation that is rather smooth and convincing, on a very humanlike face with many degrees of freedom. It also has a small number of fluid hand gestures. It is not at all clear that these are being autonomously generated during its interactions, rather than being driven as if it were a puppet. Its head movement is rather jerky, and it does not make good use of glance—where it looks—during conversations, especially when it is in a three-way conversation.

The Sophia body is normally concealed by clothing, and the recent addition of legs is well behind the state of the art in legged locomotion. It is said to implement some emotion recognition, and to have some conversational learning abilities, but these are not evidenced by the videos of the robot in action. It seems likely that its behavior is wholly or partly prescripted in some of the interviews. Sophia has much more to do with theater, or a Barnum circus attraction, than with any innovative AI.

The problem is that the novelty effect and the limits of short and tightly controlled interactions give us a false impression of what is going on. Instinctive reactions to the nicely animated face of an attractive young woman, highly visible from some of the male interviewers, lead to attributions of sentience and consciousness that have no basis whatsoever in any robots, never mind in this one.

What does this tell us about the issue of robot rights, a current topic of speculation, if mainly by people outside robotics? Let's think about this in a concrete and nonspeculative fashion. In previous chapters, we have explained how robots are constructed and what these machines can and cannot do. The key word here is *machines*. Robots are human artifacts as much as washing machines, thermostats, and the clockwork automata discussed in our introduction. We do not discuss extending rights to any of those. Why? Maybe we argue that washing machines do not move around. Okay, so how about cranes, or fly-by-wire aircraft? Or a vacuum cleaner robot? Should any of those have rights?

Perhaps we feel that these examples do not sense their environment, though, as we have pointed out, thermostats do sense their environment as well as acting on it. So do vacuum cleaner robots. These days, even industrial robot arms have some sensors. However, Saudi Arabia does not appear to be handing out citizenship to industrial robot arms or robot vacuum cleaners. At root, the argument depends on our instinctive reactions to machines in humanoid form. A robot may look somewhat human. It may have some natural language interaction abilities, even if its language use is akin to a parrot or other mimicking bird and it does not understand what it says. We still feel we should attribute degrees of humanness to such machines that they really do not have, and so we worry about rights.

The straight answer to the question posed in this chapter—could robots have morals?—is an emphatic *no*. People have morals; machines do not. However, the people who design and build the machines certainly should have morals and, moreover, responsibility for what their machines do or do not do. By suggesting that a robot should have rights or morals, we attribute responsibility to the machine rather than its makers.

We are already familiar with this deflection of responsibility through the increasing automation of information systems. For some reason, nobody agitates for software-based decision systems to be given *rights*. "The computer said no" is the tenor of many conversations with human customer services, assuming you can reach a human at all. This response is bad enough when the system is making a decision to cancel your credit card, but a good deal worse if it is an autonomous car that just ran over a pedestrian.[3]

This is the point at which Asimov's Three Laws of Robotics often come into the discussion. The Three Laws were introduced in a series of science fiction short stories by Isaac Asimov, appearing between 1940 and 1950 in science fiction magazines in the United States. They were later published in book form as *I, Robot*, and still later were the supposed inspiration behind a 2004 film of the same name. In the stories, the laws were engineered into the positronic brains of the fictional humanlike robots. The laws stated:

1. A robot may not injure a human being or, through inaction, allow a human being to come to harm.
2. A robot must obey the orders given it by human beings except where such orders would conflict with the First Law.
3. A robot must protect its own existence as long as such protection does not conflict with the First or Second Laws.

These laws sound admirably succinct and to the point. But what people often overlook is that the dramatic mainspring of most of Asimov's robot stories depended on the ways the laws were ambiguous, conflicting, or inadequate. This was true even though his fictional robots were idealized versions, minus any current real-life mobility, sensing, reasoning, and communication deficits.

Do we think the robot vacuum cleaner that ran over the dog excrement and spread it around a whole apartment "harmed" humans in violation of the First Law? Nobody died, but on the other hand, dog excrement is a source of disease, especially for children, and cleaning the apartment must have taken time, energy, and probably money. Is that *harm*?

Modern human societies have extensive legal codes precisely because simplistic formulations do not work in real-world cases. See, for example, discussions about how the sixth of Christianity's Ten Commandments, "Thou shalt not kill," plays out or should play out in modern society. Asimov's brief First Law tries to cover everything that legal codes enact, from murder to slander via affray, not to mention industrial health and safety and civil negligence. More than this, even extensive legal codes are not enough, or we would be able to carry out justice with a computer. We also require human judges, attorneys, and juries to interpret the law for specific cases. So we should not confuse interesting fictional ideas with real-world robotics.

If we cannot engineer "generic" ethics into a robot, we can require ethical conduct from the designers and manufacturers of robots. They should certainly consider specific ethical issues in what they design and build. While public concerns are mostly disproportionate to what robots can achieve now or in the foreseeable future, these concerns have prompted several high-level public bodies to work on ethical issues.

In the United States, the Institute of Electrical and Electronics Engineers (IEEE), a leading professional body, runs the IEEE Global Initiative on Ethics of Autonomous and Intelligent Systems, which is developing a variety of resources that aim "to ensure every stakeholder involved in the design and development of autonomous and intelligent systems is educated, trained, and empowered to prioritize ethical considerations so that these technologies are advanced for the benefit of humanity."[4] Meanwhile the European Union has been working on "Ethics Guidelines for Trustworthy Artificial Intelligence."[5] These cover the whole of AI, not just robotics, and have involved a committee of experts, as well as an open consultation.

So what do such ethics guidelines cover, exactly? The criteria go beyond harm to the issues of human autonomy, empowerment, and deceit. A list developed by researchers in the United Kingdom can be summarized as follows:[6]

1. Robots should not be designed solely or primarily to kill or harm humans except in the interests of national security.
2. Humans, not robots, are responsible agents. Robots are tools that should be designed and operated to comply with existing laws, including privacy.
3. Robots are products and should be designed in ways that ensure their safety and security.
4. Robots are manufactured artifacts; the illusion of emotions and intent should not be used to exploit vulnerable users. It should always be possible to tell a robot from a human.
5. It should always be possible to find out who is legally responsible for a robot.

The fourth item in the list was the one at issue with Sophia, where the difference between a robot and a human was deliberately confused. Sophia is not the only such example: news reporting about robots regularly creates such confusions, so much so that when the same experts added "seven high-level messages" to the foregoing points, the seventh read "When we see erroneous accounts in the press, we commit to take the time to contact the reporting journalists."

In the case of Sophia, the deception actually came from the producers of the robot, not just from reporters. Another example from a year later, in 2018, can be seen in news reports that a robot had been invited to "give evidence" at a Parliamentary Select Committee on Education. In an idea originating with a university marketing department, the robot was preprogrammed with answers to four questions and was presented to the committee, who then asked the questions and received the preauthored answers.[7] The only difference between this robot and an MP3 player was that the recording was surrounded by a somewhat humanoid robot body and delivered via a text-to-speech system. This is not, needless to say, what "giving evidence" actually means.

Against the background of such stories, we should not be surprised that the public has unrealistic ideas about robot capabilities. The problem here is that the bodies creating the ethical guidelines have no power to enforce them against organizations unable to resist the temptation to hype their robots. And no news organization ever lost customers by repeating such assertions.

Researchers necessarily take ethical principles more seriously, since they are required to clear experiments with ethics committees. Researchers face grave professional consequences if they are found to have violated a clearance. In the few instances where robots have made it into longer-term use, the deficiencies in what robots can actually do remove any possible confusion with humans. However, the example of the Paro seal—and other robot pets following its example—does show that vulnerable users can treat such robots as if they were real animals, even when people are told they are robots. Is this creating an unethical emotional dependency?

Let's return to the issue of harm, and the first principle in the list. Notice that it was qualified by "except in the interests of national security," which in practice is an enormous qualification indeed. Just as nuclear scientists became deeply concerned about atomic weapons, so many robotics experts are now worried about the development of lethal autonomous robots (LARs),[8] or to give them a blunter handle, killer robots.[9]

Levels of automation and technologies taken from robotics have been applied to weapons for some time. The US cruise missile has for many years navigated autonomously, using techniques similar to those discussed in chapter 5. It combines dead reckoning, GPS information, and contour matching of the terrain it flies over. The cruise missile also has automatic target recognition, which matches locally sensed data with the target allocated to it for that mission. The issue that really worries experts is giving such devices the power to autonomously select targets, as well. Imagine a swarm of autonomous drones permanently flying above a city, deciding whom to kill and then killing them.

First, we have seen that sensing technologies are not wholly reliable. Consider facial recognition. Success rates are very good if you are searching a database and matching a photo, and still good if you are close to and facing a screen—hence their use in mobile phones. Under real-world conditions, recognition rates drop sharply. In 2019 an evaluation of the technology used by a UK police force in public places showed that of forty-two "suspect matches," only eight were correct—an extremely high rate of false positives.[10] Face recognition is also known to be less accurate on darker faces and on women, because systems have been trained on data containing an unbalanced number of white males.[11] Erroneous targeting is rather likely.

Second, no autonomous system can accurately determine the risk of disproportionate harm—killing a lot of untargeted civilians, for example. This is one of many reasons why decisions to exercise lethal force require human judgment. Military drones are operated by people because the operator must receive an explicit command from a person who has the current facts before deploying the drone. It is true that concern for civilians around a military target is not always exercised, but at least the people responsible for the decision can be identified. This lack of responsibility puts the user of autonomous weapons in violation of international law: the Geneva Conventions ban the use of disproportionate force.[12]

Because a number of countries have made strong moves toward using wholly autonomous weapons, with the risk of spurring a technology-driven arms race, a growing response seeks international accords to control them. The United Nations has held four meetings under the Convention on Certain Conventional Weapons (CCW) to discuss the possibility of a treaty banning autonomous weapons. Participants agree on the need for "meaningful human control" of such weapons, but not on how this would work, or whether there should be an international law or weaker guidelines. In a climate of growing nationalisms and disregard for international treaties, this work takes on a new urgency.

Moving from killer robots to sex robots may seem perverse (if that is the right word), but both military and sex industries have been early adopters of new technologies. As we pointed out in our introduction, the story of Pygmalion and Galatea concerns a sort of sexbot. Sexbots also generate a great deal of ethical debate. Some argue that since blow-up sex dolls are legal, then sexbots should be too; others argue that the production and use of sexbots involve dangers that require control over them or even laws against them.[13]

The application of AI technologies to sexbots is still much less advanced than it is to weaponry, not least because it is easier to blow people up from a distance than to engage in intimate physical contact with them. Sexbots also require convincing humanlikeness, which, as we have seen, is a difficult target once you go beyond static external appearance. It is not so difficult to add limited chatbot capabilities, but no product today can walk, and most have posable rather than independently actuated bodies. Few discussions address the personal health and safety issues of an automatically actuated sexbot skeleton.[14]

Arguments for sexbots revolve around reducing human prostitution, making sex more accessible to people who have psychosocial or physical problems, dealing with loneliness, or serving as sex therapy devices. Arguments against are based on various points in the list of ethical concerns presented earlier. Since the larger potential market seems to be for men, would it increase the objectification and commodification of women still very present in many societies?[15] Would it drive women out of sex work by undercutting them in sexbot brothels, and if it did, would that be a good thing? On the other hand, how could a machine offer the real-life backstory that a prostitute's customers often seek?[16] Never mind provide any real companionship. Doesn't a sexbot entirely depend on confusing vulnerable users about the difference between a human and a robot? Rather than acting as therapy, might sexbots reinforce disorders such as pedophilia? A sexbot modeled on a child would not be a good thing at all. Presumably, too, a sexbot might well have an internet link that would render it vulnerable to hacking and raise privacy concerns.

Many of the arguments on both sides still have an element of the speculative and are influenced by the way film treats robots and sex, a popular theme in science fiction treatments. As we saw in chapter 11, robots today are not in general up to long-term autonomy in everyday human environments, sexual or otherwise. Sexbot vendors—currently small in number—are motivated to hype their products for sales reasons. It is easy to overestimate the capabilities these products offer now or will offer anytime soon.

Other than the unrealistic fear that robots might take over the world or replace humans as a species, the implications for employment are what bother people most about robots. This fear has been fueled by a series of reports suggesting that anywhere from 15 percent to 55 percent of jobs will be affected in the next fifteen to twenty years. However, just as issues of robot ethics are bedeviled by slurring the differences between robots and humans, so discussion of the possible impacts of robots on jobs is confused by including things that are not robots at all. As we saw in chapter 12, a chatbot is not a robot. An automated information-processing system running on the internet is also not a robot.[17] Though many reports have titles such as "Will Robots and AI Take Your Job?," nearly all of them are actually discussing computer-based automation in general.[18]

The futurologist Roy Amara once pointed out that "we tend to overestimate the impact of a new technology in the short run, but we underestimate it in the long run."[19]

Automation is not a new social process but has been under way at least since the 1850s and mass industrialization. Throughout history, many jobs have been replaced by automated processes, but many new jobs have appeared, too.

Computers became available in the mid-1950s, the internet was invented in the 1960s, and the first web browser appeared in 1990. Different categories of information-based jobs were automated along that timeline: for example, payroll clerks, then typing pools, and most recently intermediary occupations like travel agents and real estate agents, along with brick-and-mortar retail.

The impact has been cumulative as technologies have come together with the addition of smartphones, the most recent change. In the most recent period, print media and television have been challenged by digital technologies, but this could only happen once digital distribution was widespread and data communication networks could routinely support streaming in real time. A single technology often lacks impact until other technologies support it. The spread and development of data communication technologies have been the key enablers. One could make the case that the biggest changes taking place now have nothing to do with robots but are caused by the automation of many more information-based jobs. This results in front-office job losses, but also some new jobs, from web design to the back-office support for web sales.

Robot sales are still dominated by industrial robots, designed to work in a factory.[20] These were the first to be developed. The first patent for an industrial robot dates back to 1954, and they were first installed in a car factory in 1962. It was not until 1975 that commercial robots were entirely electrically driven and included microprocessors, and sales did not really take off until the 1980s. Giving robots sensors so that they can respond in a limited way to the environment around them is even more recent.

The need to organize the whole factory around industrial robots made them extremely expensive to install in existing plants, so it is not surprising that the exponential growth of the last ten years has largely been driven by sales in Asian countries, which remain the largest market. These are countries that are still in the process of industrializing and can start with the latest technology. Thirty percent of industrial robots are used in vehicle manufacturing, and the explosive growth of vehicle sales in Asian countries, especially in China, depends on new factories with new robots. The next biggest market—25 percent—lies in the electrical and electronics industries. Job losses to industrial robots in manufacturing are more related to international competition than to robots replacing workers in existing industrialized countries.

The cobots discussed in chapter 9, which work alongside human operatives in a factory, are a recent introduction. They are more attractive to small and medium factories that cannot justify the costs of reengineering the factory for standard industrial robots but need to increase productivity to compete. Sales are low compared to mainstream industrial robots but have a high growth rate: from $400 million in 2017 to $600 million in 2018.[21]

In service robotics, by far the biggest growth has occurred in logistics, which includes automated warehouses.[22] This probably relates to the growth of internet consumer sales and the warehouses needed to support them. It is likely that laboring jobs in warehouses are being affected, but also that more warehouses are being built, as well. Higher-skilled workers are needed to support the robots in use. The next biggest market in service robotics is for inspection robots. Automation is also expanding in food and drink factories, though in retail, kitchen robots, such as those doing burger flipping, are expensive and still fallible, needing human support. Personal robots are substantially robot vacuum cleaners, with a smaller number of robot grass cutters.

Advanced economies support only so many manufacturing jobs, so where do the fears of much more substantial job losses come from? Many recent reports seem to overestimate the maturity of robot technologies outside manufacturing. It is true that if all private cars, trucks, and public transport vehicles were replaced over a short time by autonomous vehicles, a large number of driving jobs would be lost. As we saw, however, autonomous vehicles are not as easy to deploy as the companies involved had assumed. Reengineering roads, especially in cities, is not quick, easy, or in some cases politically acceptable. The intersection with other pressures on technology, most of all responses to climate change, makes it difficult to judge what is likely to happen. Twenty years seems like the short term here.

Other jobs turn out to be less easy to automate than expected. Autonomous cleaning machines of various sizes have been feasible for at least thirty years, but the only ones in widespread use—and even then still in a minority of homes—are autonomous vacuum cleaners. Cleaning in offices involves complex (for a robot) navigation and sensible judgments about what needs to be cleaned and how. The people who work in cleaning have low status and are badly paid. Office cleaning robots are expensive and have limited functionality. These are all reasons why so much cleaning is still carried out by people, often using nonautonomous machines.

Tables of jobs vulnerable to robot automation often appear to overgeneralize. Just as robot vacuum cleaners do not substitute for human cleaners, so grass-cutting robots do not make grounds and maintenance staff redundant, though such machines change what people do. A prediction that US numbers in this area will drop from one million to fifty thousand in ten to twenty years seems extremely hard to justify.[23]

So in what areas are more intelligent robots likely to appear other than in military and sex industry applications? That robot capabilities really need to be specifically tailored to an application and an environment makes it most likely that robots will gradually populate specific niches. The Paro seal offers an example of a niche product related to therapy and health, and we are likely to see more of these. Hazardous environments may produce compelling arguments for greater robot autonomy, so autonomous underwater vehicles are a possible growth area, along with semiautonomous search and rescue platforms for disaster relief. The growth of cheap drones, subject to tricky regulation issues still being addressed, may well produce their own set of niche applications. Inspection applications from traffic jams to the state of house roofs will become feasible once drones have longer flight times. Policing applications are also highly likely, since monitoring a protest with a drone is probably cheaper than with a helicopter. Despite extensive CCTV coverage in many cities, other drone surveillance applications seem likely too, probably linked to the camera network.

It is no surprise, however, that press coverage may mislead the public about the capabilities of robots in new domains, as we have seen. In 2018 a UK research team ran a one-week experiment to test the functionality of a social robot in a small chain of corner shops. The researchers were taking part in a series made by a broadcast television company. To their dismay, their findings appeared as a jokey twenty-second item titled "Robot Shop Assistant Sacked after One Week." The story subsequently circulated all around the world: fake news with a vengeance.[24]

Social robots represent a new area of application, and a series of companies have tried to fill this niche. Unfortunately, few of them have broken into markets outside research, with many going out of business.[25] End users require greater functionality and better long-term performance from social robots than currently feasible—more than chatbot capabilities with some added expressive behavior. The advent of voice assistants as internet interfaces means that a social robot has to turn its physical embodiment and mobility into a concrete advantage for users. This has not yet happened, though future research may overcome the problem.

We know after more than 250 years of industrial development that automation does affect jobs. The biggest dislocations are caused by rapid changes, as with the introduction of the factory system. However, we also know that new goods and services result from automation, as well. The balance between the two is much less simple to assess than alarmist press reports suggest.

What is absent from frenzied speculation about the possible impact of robots (or of automated information systems) is any consideration of the political aspects that in the end determine how damaging change is for those on the wrong end of it. If jobs are lost in coal mining, why are workers not retrained in new skills? If jobs are lost in manufacturing, why are more resources not invested in caregiving jobs, on the up in many countries because of an aging population? If retail workers are hit by internet sales, why must the extra delivery jobs that result be insecure zero-hours ones? If automation raises productivity, why does this not result in a four-day working week rather than growing income inequality and casualization of employment?

These questions are not technical but political. Applying AI and robot technologies to socially negative domains like killer robots and authoritarian surveillance is a political decision. Using automated information systems as a shield against human responsibility for their decisions is also a political decision. They are decisions that should concern all citizens.

Finally, let us return to the issue of what we think robots are. Throughout the book, we have argued strongly that they are machines, engineered artifacts, and not a sort of living thing. Most robot bodies are made of metal and motors, powered by batteries, and driven by software developed by human programmers running on computers. At their simplest, robots are metal boxes on wheels with a computer inside them.

This is fundamentally different from the way living things are constructed. A human is not a computerlike brain embedded in a fleshlike box. This is a convenient metaphor, but it is not much more accurate than earlier metaphors that saw brains as hydraulic systems or telephone exchanges. A body is a complex of interlocking electrochemical processes, where internal and external behavior emerges from self-maintaining dynamic interactions.[26] For this reason, simply giving a robot better software is not likely to move it toward being more humanlike. The dynamic complexity of the human body is a major reason why we still do not understand quite a lot about how it works.

The brain is an extension of a nervous system that permeates the whole body and within which a great deal of activity is locally regulated, without the brain's intervention. The nervous system is not the only network, either; the circulatory system and the endocrine system also permeate the body. Many bodily processes, both local and distributed, support reproduction through the sharing of male and female genetic information. This is then used by processes in the female body to produce a new child. The body's energy is supplied by a set of electrochemical processes that take in oxygen and water, break down food, and excrete waste.

A robot body does also contain electrochemical processes, for example, corrosion. These are not integrated together, because the robot is composed of many discrete elements, each of which has had to be manufactured. Where components in a robot are linked, it is through physical contact, involving screws or other fixings, components such as gears, occasionally pipes containing hydraulics or compressed air, and usually electrical circuits. Behind a human child are a couple of parents. Behind a robot is a worldwide manufacturing complex, which itself is the product of earlier manufacturing, dating back to the Industrial Revolution. For a robot to be able to control the production of more robots, it would have to control much of this complex, all the way back to mining raw materials.

Roboticists soon gain a deep respect for the capabilities of living things as they struggle to extract aspects that can be engineered into their robots. We are making progress, but it is slow, because the problems being tackled are monumentally hard to solve, whether for hardware or for software. As we have examined how robots move, navigate, grasp, use AI, learn, model emotions, cooperate, and try to perform in social environments, we have seen just how difficult it is to replicate more than a small fraction of human abilities. We really cannot build a robot that is just like us.

Nor is there any sign of some so far unthought-of breakthrough that will make robotics issues suddenly simpler and allow the rapid robot advances that speculative futurology seems to assume. That is not to say that robotics is pointless, any more than medicine is pointless because of our incomplete understanding of our bodies. Forget the hype, and the unrealistic speculations about so-called new species or superhuman abilities, and let's work together as roboticists and as citizens to apply this fascinating and challenging technology where it will do some good.

ACKNOWLEDGMENTS

We acknowledge the help and support of the robotics research community at large in writing this book, though any errors are those of the authors alone. Specific thanks for the use of photos to the partners of the EU project LIREC and the UKRI project SoCoRo, to the Edinburgh Centre for Robotics, to James Law and the Sheffield Centre for Robotics, to Andy Wallace and Phil Bartie of Heriot-Watt University, and Alan Law of the University of Stirling, all in the United Kingdom. Also to Michael Walters of University of Hertfordshire. Tim Perkins, Sheila Perkins, Greg Michaelson, and Verena Reiter all read some or all of the book and made helpful comments.

NOTES

INTRODUCTION

1. Western European attitudes to robots have become more negative over the last ten years. See T. Gnambs and M. Appel, "Are Robots Becoming Unpopular? Changes in Attitudes towards Autonomous Robotic Systems in Europe," *Computers in Human Behavior* 93 (2019): 53–61.

2. Marvin Minsky discussed "suitcase words" in *The Emotion Machine* (Simon & Schuster, 2006). For a synopsis, see Rodney Brooks, "The Seven Deadly Sins of AI Predictions," *MIT Technology Review*, October 6, 2017, https://www.technologyreview.com/s/609048/the-seven-deadly-sins-of-ai-predictions (accessed November 20, 2020).

CHAPTER 1

1. For this myth and other similar ones, see Adrienne Mayor, *Gods and Robots* (Princeton, NJ: Princeton University Press, 2018). In Ovid's story, the statue had no name; she became Galetea in later versions.

2. The *Oxford English Dictionary* cites the earliest use of the term *android* as *androides* in Ephraim Chambers's *Cyclopaedia* of 1728, used in connection with an automaton said to have been built by Saint Albertus Magnus.

3. See especially the "humanlike" robots developed in Japan. Intelligent graphical characters are also mostly represented as women, and home conversational interfaces (Alexa, Siri, Google Home) are usually given the voice of a young woman.

4. Mayor, *Gods and Robots*, 90–95.

5. One of the authors discusses the case of the so-called robot artist Ai-Da: Ruth Aylett, "Ai-Da: A Robot Picasso or Smoke and Mirrors?" Medium.com, July 13, 2019, https://medium.com/@r.s.aylett/ai-da-a-robot-picasso-or-smoke-and-mirrors-a77d4464dd92 (accessed November 20, 2020).

6. Ktesibios is discussed in E. A. Truitt, *Medieval Robots: Mechanism, Magic, Nature and Art* (University of Pennsylvania Press, 2015), 4, 156n8.

7. Truitt, *Medieval Robots*, 31–32. The singing birds at the court of Byzantium passed into popular history; see, e.g., W. B. Yeats's poems "Byzantium" and "Sailing to Byzantium."

8. Mayor, *Gods and Robots*, 198–199.

9. Truitt, *Medieval Robots*, 4.

10. Mayor, *Gods and Robots*, 200–201.

11. Truitt, *Medieval Robots*, 122–137.

12. Tony Freeth, Yanis Bitsakis, Xenophon Moussas, John H. Seiradakis, A. Tselikas, H. Mangou, M. Zafeiropoulou, et al., "Decoding the Ancient Greek Astronomical Calculator Known as the Antikythera Mechanism," *Nature* 444, no. 7119 (November 2006): 587–591.

13. Truitt, *Medieval Robots*, 147.

14. Church automata are discussed in Jessica Riskin, "Machines in the Garden," *Republics of Letters* 1, no. 2 (April 30, 2010), https://arcade.stanford.edu/rofl/machines-garden (accessed November 20, 2020).

15. Renato and Franco Zamberlan, "The St. Mark's Clock, Venice," *Horological Journal*, January 2001, 11–14.

16. The National Museum of American History has a comprehensive video of the monk automaton's movements: https://www.youtube.com/watch?v=kie96iRTq5M (accessed November 20, 2020).

17. Cartesian dualism has been strongly challenged both by materialist philosophers like Daniel Dennett and by neurophysiologists like Antonio Damasio. They argue that mind is a material property of the

brain, and the brain is an integrated part of the body, not a separate control center.

18. The UK project AIKON-II (2008–2012) investigated how a robot arm could draw a portrait from a camera picture; see https://sites.google.com/site/aikonproject/Home/aikon-ii (accessed November 20, 2020). This represents an advance on the Draughtsman, whose four pictures were predetermined; it used graphics algorithms to encode some artistic knowledge about how the drawing should be executed.

19. Basile Bouchon, a silk worker in Lyon, invented a paper-tape-directed loom in 1725. The automation of looms culminated in the Jacquard loom in 1804.

20. DARPA ran a challenge in 2015. Robots had to deal with an obstacle course that tested their capabilities for search and rescue operations. Tasks included walking on unstable surfaces, climbing stairs, opening doors, getting out of a car, and closing a valve. The robots viewable in some outtakes from the finals were not, however, autonomous but remotely operated most of the time; see "Robots Falling Down at DARPA Robotics Challenge," *IEEE Spectrum*, video posted to YouTube, June 13, 2019, https://www.youtube.com/watch?v=xb93Z0QItVI (accessed November 20, 2020).

21. Some researchers in human-robot interaction think the Wizard of Oz technique is overused; see Laurel Riek, "Wizard of Oz Studies in HRI: A Systematic Review and New Reporting Guidelines," *Journal of Human-Robot Interaction* 1, no. 1 (2012): 119–136, https://doi.org/10.5898/JHRI.1.1.Riek.

22. Boston Dynamics, an internationally leading robot engineering company, shows many videos of its robots—though most seem to be teleoperated. In one video, the robot navigates autonomously, though only in a static environment it has already been taken around: "Spot Autonomous Navigation," YouTube, May 10, 2018, https://www.youtube.com/watch?v=Ve9kWX_KXus (accessed November 20, 2020). The Boston Dynamics videos became so popular that an organization called Corridor produced a spoof in which abused robots eventually fight back: "Boston Dynamics: The Robot Fight Back (Corridor Parody)," YouTube, June 16, 2019, https://www.youtube.com/watch?v=rW9WmA5okpE (accessed November 20, 2020). Large numbers of people were initially deceived into thinking this was footage of actual robots.

23. For a brief discussion of this issue, see Christopher Mims, "Why Japanese Love Robots (and Americans Fear Them)," *MIT Technology Review*, October 20, 2010, https://www.technologyreview.com/s/421187/why-japanese-love-robots-and-americans-fear-them. A longer discussion of Japanese attitudes and a possible connection with Shintoism appears in a paper by Naho Kitano, "Animism, Rinri, Modernization: The Base of Japanese Robotics," in *Proceedings 2007 IEEE International Conference on Robotics and Automation*, vol. 7 (ICRA, April 2007), 10–14.

CHAPTER 2

1. "$10 Million Awarded to Family of U.S. Plant Worker Killed by Robot," *Ottawa Citizen*, August 11, 1983, 14.

2. For an argument that a thermostat is an intelligent agent, see Stuart J. Russell and Peter Norvig, *Artificial Intelligence: A Modern Approach*, 2nd ed. (Upper Saddle River, NJ: Prentice Hall, 2003), chap. 2; for an argument that it is not, see the Consortium on Cognitive Science Instruction (CCSI) online course "Introduction to Intelligent Agents," http://www.mind.ilstu.edu/curriculum/ants_nasa/intelligent_agents.php (accessed November 20, 2020).

3. Gibson originated the idea in 1966 but explored it in greater detail in *The Ecological Approach to Visual Perception* (Boston: Houghton Mifflin Harcourt, 1979).

4. Not just true of robots. The "computers are social actors" theory has argued since 1994 that we tend to see all computers, not just those driving a robot, as social actors. This means applying social norms such as politeness and sociable interaction and involves attributing personality and an independent inner life. These ideas are most associated with Clifford Nass and his team; see Clifford Nass, Jonathan Steuer, and Ellen R. Tauber, "Computers Are Social Actors," in *Proceedings of the SIGCHI Conference on Human Factors in Computing Systems* (ACM, 1994), 72–78.

5. Observed in a series of experiments carried out in 2013 at MIT by Cynthia Breazeal, a pioneer of social robotics. This represents one of a number of instances where human participants thought the robot was being condescending, though the robot had no means of choosing such behavior. C. Breazeal, N. DePalma, J. Orkin, S. Chernova, and M.

Jung, "Crowdsourcing Human-Robot Interaction: New Methods and System Evaluation in a Public Environment," *Journal of Human-Robot Interaction* 2, no. 1 (2013): 82–111.

6. In the 2006 experiments with pictures of eyes and flowers in a real-world setting, researchers found coffee money contribution levels always increased with the transition from flowers to eyes, and decreased with the transition from eyes to flowers. M. Bateson, D. Nettle, and G. Roberts, "Cues of Being Watched Enhance Cooperation in a Real-World Setting," *Biology Letters* 2, no. 3 (2006): 412–414. When the idea was extended to littering behavior, half the amount of littering occurred when eyes were shown compared to flowers. M. Ernest-Jones, D. Nettle, and M. Bateson, "Effects of Eye Images on Everyday Cooperative Behavior: A Field Experiment," *Evolution and Human Behavior* 32, no. 3 (2011): 172–178.

7. E. Broadbent, R. Tamagawa, N. Kerse, B. Knock, A. Patience, and B. MacDonald, "Retirement Home Staff and Residents' Preferences for Healthcare Robots," in *RO-MAN 2009: The 18th IEEE International Symposium on Robot and Human Interactive Communication* (IEEE, September 2009), 645–650.

8. Work carried out with a commercially available robot toy, the Pleo dinosaur, was inspired by the way people already personalize their laptops and mobile phones. Items such pajamas and a bracelet were used to change its behaviors. Y. Fernaeus and M. Jacobsson, "Comics, Robots, Fashion and Programming: Outlining the Concept of Act-Dresses," in *Proceedings of the 3rd International Conference on Tangible and Embedded Interaction* (ACM, February 2009), 3–8).

9. For this experiment at the University of Hertfordshire, see M. L. Walters, K. L. Koay, D. S. Syrdal, K. Dautenhahn, and R. Te Boekhorst, "Preferences and Perceptions of Robot Appearance and Embodiment in Human-Robot Interaction Trials," in *Proceedings of New Frontiers in Human-Robot Interaction 2009*, https://uhra.herts.ac.uk/handle/2299/9642 (accessed November 20, 2020).

10. Folding origami robots are aimed at democratizing access to customized robots for industrial, home, and educational use by making them easy to reconfigure. C. D. Onal, M. T. Tolley, R. J. Wood, and D. Rus, "Origami-Inspired Printed Robots," *IEEE/ASME Transactions on Mechatronics* 20, no. 5 (2014): 2214–2221.

11. The uncanny valley was a hypothesis, but a hugely influential one in subsequent robot and graphic character design. In Japanese, see Masahiro Mori, *The Buddha in the Robot* (Tokyo: Kosei Publishing, 1981); an English translation was published by Kosei Shuppan-Sha in 1992.

12. For experiments investigating what produces an uncanny valley reaction, see Maya B. Mathur and David B. Reichling, "Navigating a Social World with Robot Partners: A Quantitative Cartography of the Uncanny Valley," *Cognition* 146 (January 2016): 22–32, https://doi.org/10.1016/j.cognition.2015.09.008.

13. For Kismet in action, along with an explanation of its behavior, see Radhika Khandelwal, "Kismet the AI Robot," YouTube, October 8, 2014, https://www.youtube.com/watch?v=NpbCPNoLqd0 (accessed November 20, 2020).

14. The animators' bible is *The Illusion of Life: Disney Animation*, by two long-standing Disney animators, Frank Thomas and Ollie Johnston. It includes the lessons they learned over their long careers in giving animated characters believable personalities.

CHAPTER 3

1. For the ideas being used for legged robots, see Q. D. Wu, C. J. Liu, J. Q. Zhang, and Q. J. Chen, "Survey of Locomotion Control of Legged Robots Inspired by Biological Concept," *Science in China Series F: Information Science* 52, no. 10 (2009): 1715–1729.

2. Boston Dynamics trialed its Big Dog robot for the US Army starting in 2005. See the IEEE robot catalog at https://robots.ieee.org/robots/bigdog (accessed November 20, 2020). The US Army did not adopt Big Dog in the end; the robot's internal combustion engine, used to give it a good range, was seen as too noisy for the sort of stealthy infantry operations the army was targeting.

After Big Dog, Boston Dynamics produced smaller doglike robots called Spot. For an account of the close-to-production version, see TechCrunch, "Marc Raibert Shows Off a Close-to-Production Spot Mini," YouTube, April 22, 2019, https://www.youtube.com/watch?v=iBt2aTjCNmI (accessed November 20, 2020). Spot runs for about ninety minutes between recharges and is now for sale at $74,500. For a useful corrective to the expectations that some of the Spot videos have aroused, see Matt Simon, "Spot, the Internet's Wildest 4-Legged Robot,

Is Finally Here," *Wired*, September 24, 2019, https://www.wired.com/story/spot-boston-dynamics (accessed November 20, 2020).

3. For a short account of an experiment to roboticize actual cockroaches, see Ian Sample, "Cockroach Robots? Not Nightmare Fantasy but Science Lab Reality," *The Guardian*, March 3, 2015, https://www.theguardian.com/science/2015/mar/04/cockroach-robots-not-nightmare-fantasy-but-science-lab-reality (accessed November 20, 2020).

4. For footage of the original hopping robot, see Plastic Pals, "Robots from MIT's Leg Lab," YouTube, October 31, 2011, https://www.youtube.com/watch?v=XFXj81mvInc (accessed November 20, 2020).

5. Hydraulic fluid leaks after falls at the DARPA challenge in 2015: Evan Ackerman and Erico Guizzo, "DARPA Robotics Challenge: Amazing Moments, Lessons Learned, and What's Next," *IEEE Spectrum*, June 11, 2015, https://spectrum.ieee.org/automaton/robotics/humanoids/darpa-robotics-challenge-amazing-moments-lessons-learned-whats-next (accessed November 20, 2020).

6. A beautiful video of the Boston Dynamics Atlas biped doing gymnastics: Engadget, "Boston Dynamics' Atlas Robot Now Does Gymnastics, Too," YouTube, September 25, 2019, https://www.youtube.com/watch?v=kq6mJOktIvM (accessed November 20, 2020). The more recent Honda ASIMO doing autonomous obstacle avoidance: Auto Channel, "Honda Unveils All-New ASIMO with Significant Advancements," YouTube, November 12, 2011, https://www.youtube.com/watch?v=yND4k4NM0qU (accessed November 20, 2020).

7. Carnegie Mellon roboticists reflect on the performance of their two teams in the 2015 DARPA challenge: C. G. Atkeson et al., "What Happened at the DARPA Robotics Challenge, and Why?" https://www.cs.cmu.edu/~cga/drc/jfr-what.pdf (accessed November 20, 2020).

8. Table of energy densities: https://en.wikipedia.org/wiki/Energy_density (accessed November 20, 2020).

9. G. Gabrieli and T. von Karman, "What Price Speed," *Mechanical Engineering* 72, no. 10 (1950): 775–781.

10. Discussion of energy efficiency in robots: Navvab Kashiri et al., "An Overview on Principles for Energy Efficient Robot Locomotion," *Frontiers in Robotics and AI*, December 11, 2018, https://www.frontiersin.org/articles/10.3389/frobt.2018.00129/full#note2 (accessed November 20, 2020).

11. Work on sugar batteries: Sebastian Anthony, "Sugar-Powered Bio-battery Has 10 Times the Energy Storage of Lithium: Your Smartphone Might Soon Run on Enzymes," *ExtremeTech*, January 21, 2014, https://www.extremetech.com/extreme/175137-sugar-powered-biobattery-has-10-times-the-energy-storage-of-lithium-your-smartphone-might-soon-run-on-enzymes (accessed November 20, 2020).

12. For recent work in bio-inspired flying robots, see Kate Baggaley, "Forget Props and Fixed Wings: New Bio-inspired Drones Mimic Birds, Bats and Bugs," NBC News, July 30, 2019, https://www.nbcnews.com/mach/science/forget-props-fixed-wings-new-bio-inspired-drones-mimic-birds-ncna1033061 (accessed November 20, 2020).

13. Smartbird overview in the IEEE robot catalog: https://robots.ieee.org/robots/smartbird (accessed November 20, 2020).

14. For the French Jessiko robot fish, aimed at entertainment, see http://www.robotswim.com/?lang=English (accessed November 20, 2020). It is small (23 cm long) and is said to be able to swim for nine hours between recharges, carried out by induction. It includes LEDs that can change color.

15. The IEEE has video footage of a snakebot that can swim as well as slither, though it is not clear whether it is being teleoperated: "Snake Bot," *IEEE Spectrum*, video posted to YouTube, December 23, 2013, https://www.youtube.com/watch?v=vCrN47cOmHQ (accessed November 20, 2020).

CHAPTER 4

1. For the story as reported in the press at the time, see Martin Wainwright, "Robot Fails to Find a Place in the Sun," *The Guardian*, June 20, 2002, https://www.theguardian.com/uk/2002/jun/20/engineering.highereducation (accessed November 20, 2020).

2. George Stratton, a pioneering US psychologist, carried out early experiments with inverting glasses as a PhD student in Leipzig, Germany, in the years around 1900. His glasses inverted both up and down and left and right. In his 1897 paper, he claimed that after a week he saw the world the right way up again. Later work—in Austria in the 1950s—suggested that people who wore the glasses did not see the world right way up but were able to adapt their behavior as if they were themselves upside down in the real world. You can watch a

classic film on the Austrian work at https://www.youtube.com/watch?v=jKUVpBJalNQ.

3. A widely used open-source library used in vision processing is OpenCV (Open Computer Vision), https://opencv.org. This holds all the classic algorithms and now also incorporates machine-learning-based approaches.

4. Robot ecosystems have been studied for a long time; for an example from researchers in Brussels more than twenty years ago, see A. Birk, "Where to Watch" (1997), Semantic Scholar, https://pdfs.semanticscholar.org/50b4/51436a7e1608fc8c72789533e49c7a88a3ca.pdf (accessed November 20, 2020).

5. For more on how the human olfactory system works, see Allison Marin (Curley), "Making Sense of Scents: Smell and the Brain," BrainFacts.org, January 27, 2015, https://www.brainfacts.org/thinking-sensing-and-behaving/smell/2015/making-sense-of-scents-smell-and-the-brain (accessed November 20, 2020).

6. For the gasbot, see V. H. Bennetts et al., "Gasbot: A Mobile Robotic Platform for Methane Leak Detection and Emission Monitoring" (2012), Semantic Scholar, https://pdfs.semanticscholar.org/8ff6/de5226e43b67c3993c63f995cb001eeada99.pdf (accessed November 20, 2020).

CHAPTER 5

1. The Kalman filter is named after a Hungarian immigrant to the United States, who became a citizen and worked along with other engineers and mathematicians to produce the filter in the late 1950s and early 1960s. For a simple illustrated explanation of the Kalman filter, see https://www.bzarg.com/p/how-a-kalman-filter-works-in-pictures (accessed November 20, 2020). Any internet search for the term will bring up the information you need to implement one yourself. See also YouTube tutorials from MatLab, the tool kit often used for implementation, https://www.youtube.com/watch?v=mwn8xhgNpFY (accessed November 20, 2020).

2. This view of robotics came from various researchers but is most associated with Rodney Brooks of MIT. Brooks first published on this subject in 1987, but the definitive paper is "Intelligence without Representation," *Artificial Intelligence* 47, nos. 1–3 (1991): 139–159.

3. Ron Arkin, another leading researcher in this field, covered the topic comprehensively. Ronald C. Arkin, *Behavior-Based Robotics* (Cambridge, MA: MIT Press, 1998).

4. For a tutorial description of SLAM, see Søren Riisgard and Rufus Blas Morten, "SLAM for Dummies: A Tutorial Approach to Simultaneous Localization and Mapping," 2005, http://citeseerx.ist.psu.edu/viewdoc/summary?doi=10.1.1.208.6289 (accessed November 20, 2020).

5. The scientists who came up with SLAM were Hugh Durrant-Whyte and John Leonard, drawing on work by three earlier scientists.

6. Wikipedia has an informative account of the DARPA Grand Challenges at https://en.wikipedia.org/wiki/DARPA_Grand_Challenge (accessed November 20, 2020).

7. Opened in 1987, the Docklands Light Railway (DLR) in London was the first driverless metro, at level 3 autonomy. Since then, many level 4 systems have been and are being installed. Some are short airport systems, and some still carry a human operator even though they are capable of running without one.

8. Since 2008, Rio Tinto has been mining iron ore with increasing levels of automation in the Pilbara region of Australia. Arid and very hot working conditions were problematic for human operators. A control center in Perth, hundreds of miles away, allows remote supervision and high-level control of operations. Autonomous trains to carry the iron ore to ports have recently been added.

9. For the IEC International Standard for urban guided transport management and command/control systems, see https://webstore.iec.ch/publication/6777 (accessed November 20, 2020).

10. The US Union of Concerned Scientists cites six levels of autonomy in a discussion at https://www.ucsusa.org/resources/self-driving-cars-101 (accessed November 20, 2020).

11. For the preliminary report into the accident by the US National Transportation Safety Board, see https://www.ntsb.gov/investigations/AccidentReports/Reports/HWY18MH010-prelim.pdf (accessed November 20, 2020).

12. For a more cautious assessment from a Toyota designer—and former roboticist—see Philip E. Ross, "Q&A: The Masterminds behind Toyota's Self-Driving Cars Say AI Still Has a Way to Go," *IEEE*

Spectrum, June 29, 2020, https://spectrum.ieee.org/transportation/self-driving/qa-the-masterminds-behind-toyotas-selfdriving-cars-say-ai-still-has-a-way-to-go.

CHAPTER 6

1. Deep Fritz won a series four games to two against the world chess champion Vladimir Kramnik back in 2006. See Chessbase.com, https://en.chessbase.com/post/kramnik-vs-deep-fritz-computer-wins-match-by-4-2 (accessed November 20, 2020).

2. For discussion of part of AlphaGo's match, see Fun Call Centre, "Lee Sedol vs AlphaGo Move 37 Reactions and Analysis," YouTube, January 5, 2018, https://www.youtube.com/watch?v=HT-UZkiOLv8&t=43s (accessed November 20, 2020). It is clear that the program calculates the move, but humans make it.

3. For an account of the 2010 Small-Scale Manipulation Challenge in which robots played chess with Staunton pieces on an ordinary board, see Monica Anderson et al., "Report on the AAAI 2010 Robot Exhibition," *AI Magazine* 32, no. 3 (Fall 2011), https://aaai.org/ojs/index.php/aimagazine/article/view/2317; for the 2011 rerun, see Sonia Chernova et al., "The AAAI 2011 Robot Exhibition," *AI Magazine* 33, no. 1 (Spring 2012), https://aaai.org/ojs/index.php/aimagazine/article/view/2398 (accessed November 20, 2020).

4. This article discusses whether increasing automation in food and hospitality services will result in job losses: Alana Semuels, "Robots Will Transform Fast Food," *The Atlantic*, January–February 2018, https://www.theatlantic.com/magazine/archive/2018/01/iron-chefs/546581 (accessed November 20, 2020).

5. For a prototype of a Japanese pancake-flipping robot, see Ikinamo, "A Robot That Cooks Japanese Okonomiyaki Pancakes," YouTube, June 16, 2009, https://www.youtube.com/watch?v=nv7VUqPE8AE (accessed November 20, 2020). As usual, the technology is not the only issue in introducing robots. For an article about a deployed burger-flipping robot in California, see Mark Austin, "Flippy Gets Fired: Burger Bot Shut Down after One Day on the Job," Digitaltrends.com, March 10, 2018, https://www.digitaltrends.com/cool-tech/flippy-burger-flipping-robot-shut-down. This robot disrupted the kitchen workflow. Its pace could not be regulated according to demand, so that it worked

badly with humans, who were still needed to season the patties before cooking and add condiments and serve afterward. Its installation in 2018 was therefore paused.

6. On pneumatic muscles, see George Andrikopoulos, Stamatis Manesis, and George Nikolakopoulos, "A Survey on Applications of Pneumatic Artificial Muscles," paper presented at the 19th Mediterranean Conference on Control and Automation (MED), Corfu, Greece, June 2011, https://www.researchgate.net/profile/George_Andrikopoulos/publication/259310538_A_Survey_on_applications_of_Pneumatic_Artificial_Muscles/links/55a969f108ae481aa7f985c1/A-Survey-on-applications-of-Pneumatic-Artificial-Muscles.pdf (accessed November 20, 2020).

7. On using shape memory technology to produce artificial muscles that lengthen and shorten, see University Saarland, "Keeping a Tight Hold on Things: Robot-Mounted Vacuum Grippers Flex Their Artificial Muscles," *ScienceDaily*, March 23, 2018, https://www.sciencedaily.com/releases/2018/03/180323090956.htm (accessed November 20, 2020).

8. Though now out of production, the Baxter robot was produced specifically as an industrial robot that would be safe for humans working around it. Matt Simon, "A Long Goodbye to Baxter, a Gentle Giant among Robots," *Wired*, October 8, 2018, https://www.wired.com/story/a-long-goodbye-to-baxter-a-gentle-giant-among-robots (accessed November 20, 2020).

9. For work at the University of Houston on soft sensors for a robot hand, see Jeannie Kever, "Artificial 'Skin' Gives Robot Hand a Sense of Touch," Phys.org, September 13, 2017, https://phys.org/news/2017-09-artificial-skin-robotic.html (accessed November 20, 2020).

10. For a project to produce a tactile skin interface for a robot phone, see https://marcteyssier.com/projects/skin-on (accessed November 20, 2020). The artificial skin provoked many uncanny valley reactions; people seem to find it "creepy."

11. For a slightly dated survey on soft robots that presents the challenges well, see Deepak Trivedi et al., "Soft Robotics: Biological Inspiration, State of the Art, and Future Research," *Applied Bionics and Biomechanics* 5, no. 3 (October 2008): 99–117, https://doi.org/10.1080/11762320802557865.

12. Researchers at Yale animated a soft toy with their elastic robot skin. William Weir, "'Robotic Skins' Turn Everyday Objects into Robots," *Yale News*, September 19, 2018, https://news.yale.edu/2018/09/19/robotic-skins-turn-everyday-objects-robots. See also J. W. Booth, D. Shah, J. C. Case, E. L. White, M. C. Yuen, O. Cyr-Choiniere, and R. Kramer-Bottiglio, "OmniSkins: Robotic Skins That Turn Inanimate Objects into Multifunctional Robots," *Science Robotics* 3, no. 22 (2018): eaat1853, https://robotics.sciencemag.org/content/3/22/eaat1853 (accessed November 20, 2020).

13. For an example of a commercially available robot hand with twenty-four degrees of freedom, see https://www.shadowrobot.com/products/dexterous-hand (accessed November 20, 2020).

14. For an article about a US research group developing a robot hand that is structurally really close to a human hand, see Evan Ackerman, "This Is the Most Amazing Biomimetic Anthropomorphic Robot Hand We've Ever Seen," *IEEE Spectrum*, February 18, 2016, https://spectrum.ieee.org/automaton/robotics/medical-robots/biomimetic-anthropomorphic-robot-hand (accessed November 20, 2020).

15. Some of the basic ideas on prosthetics are explained in B. Dellon and Y. Matsuoka, "Prosthetics, Exoskeletons, and Rehabilitation (Grand Challenges of Robotics)," *IEEE Robotics and Automation Magazine* 14, no. 1 (2007): 30–34, http://citeseerx.ist.psu.edu/viewdoc/download?doi=10.1.1.206.2443&rep=rep1&type=pdf (accessed November 20, 2020).

16. For a survey of evidence from many papers about the experience of lower-limb prosthesis users in the United States, see the National Center for Biotechnology Information (NCBI), "Lower Limb Prostheses: Measurement Instruments, Comparison of Component Effects by Subgroups, and Long-Term Outcomes," https://www.ncbi.nlm.nih.gov/books/NBK531527 (accessed November 20, 2020).

17. For an optimistic user view of modern prostheses, see Patrick Kane, "Being Bionic: How Technology Changed My Life," *The Guardian*, November 15, 2018, https://www.theguardian.com/technology/2018/nov/15/being-bionic-how-technology-transformed-my-life-prosthetic-limbs (accessed November 20, 2020).

18. For a recent discussion of some of the issues involved in exoskeletons for people with spinal injuries, see A. S. Gorgey, "Robotic

Exoskeletons: The Current Pros and Cons," *World Journal of Orthopedics* 9, no. 9 (2018): 112, https://www.ncbi.nlm.nih.gov/pmc/articles/PMC6153133 (accessed November 20, 2020).

19. For a news article about driving an exoskeleton via a brain-computer interface, see Amy Woodyatt, "Paralyzed Man Walks Using Brain-Controlled Robotic Suit," CNN, October 4, 2019, https://edition.cnn.com/2019/10/04/health/paralyzed-man-robotic-suit-intl-scli/index.html (accessed November 20, 2020).

20. For a recent overview of industrial exoskeletons, see Dan Kara, "Industrial Exoskeletons: New Systems, Improved Technologies, Increasing Adoption," *Robot Report,* December 6, 2018, https://www.therobotreport.com/industrial-exoskeletons (accessed November 20, 2020).

21. Hand of Hope is one example of several poststroke rehabilitation exoskeletons. See http://www.rehab-robotics.com/hoh; and see it in action on YouTube at https://www.youtube.com/watch?v=9Ysa7FmDWrk (accessed November 20, 2020).

CHAPTER 7

1. A November 2018 article (updated June 2019) in *Psychology Today* by the psychologist Neel Burton about definitions of intelligence suggests looking at the impact of dementia for a guide to the dimensions of intelligence; see https://www.psychologytoday.com/gb/blog/hide-and-seek/201811/what-is-intelligence (accessed November 20, 2020).

2. In *The Mismeasure of Man* (1980), Stephen Jay Gould argues against "the argument that intelligence can be meaningfully abstracted as a single number capable of ranking all people on a linear scale of intrinsic and unalterable mental worth."

3. The General Problem Solver of Herbert A. Simon, J. C. Shaw, and Allen Newell was written in an early programming language, Fortran. It articulated a theory of how humans solve problems, but it could only be applied to "well-defined" problems. An example would be how to get a fox, a goose, and a cabbage over a river in a single-passenger boat, given that an unchaperoned fox would eat the goose, and an unsupervised goose would eat the cabbage. For a discussion of the whole approach, see Simon and Newell, "Human

Problem-Solving: The State of the Theory in 1970," https://pdfs.semanticscholar.org/18ce/82b07ac84aaf30b502c93076cec2accbfcaa.pdf (accessed November 20, 2020).

4. This relates to a discussion known as *symbol grounding*. Proponents argue that no computer system can be intelligent unless it can link the symbols it manipulates to an engagement with the real world. Thus an intelligent computer has to have sensors and link them through to its symbols, or in fact be a robot.

5. The cognitive scientist Philip Johnson-Laird argued that humans do not implement deduction as a set of logical rules, but "reasoning depends on envisaging the possibilities consistent with the starting point—a perception of the world, a set of assertions, a memory, or some mixture of them." He shows that the same logical problem can be solved when given a real-world setting, but not when it is posed abstractly. See, e.g., P. N. Johnson-Laird, "Mental Models and Human Reasoning," *Proceedings of the National Academy of Sciences* 107, no. 43 (2010): 18243–18250.

6. This definition is ascribed to the AI pioneer Marvin Minsky; see a biography at https://www.britannica.com/biography/Marvin-Lee-Minsky (accessed November 20, 2020).

7. The two most influential theoretical texts of this movement were Terry Winograd and Fernando Flores, *Understanding Computers and Cognition: A New Foundation for Design* (Intellect Books, 1986); and Lucy Suchman, *Plans and Situated Actions: The Problem of Human-Machine Communication* (Cambridge University Press, 1987). Both argued against the AI consensus of the day.

8. For this definition and a discussion of possible social impacts of AI, see Joanna J. Bryson, "The Past Decade and Future of AI's Impact on Society," BBVA OpenMind, February 2019, https://www.bbvaopenmind.com/wp-content/uploads/2019/02/BBVA-OpenMind-Joanna-J-Bryson-The-Past-Decade-and-Future-of-AI-Impact-on-Society.pdf (accessed November 20, 2020).

9. Many researchers worked on robot architectures involving linked sets of behavioral reactions. One of the best known was the *subsumption architecture* of Rodney Brooks. See Brooks, "Intelligence without Representation," *Artificial Intelligence* 47, nos. 1–3 (1991): 139–159, https://courses.media.mit.edu/2002fall/mas962/MAS962/brooks%203.1.pdf (accessed November 20, 2020).

10. For Shakey the robot, the first project to apply these AI ideas to a robot, see Daniela Hernandez, "Tech Time Warp of the Week: Shakey the Robot, 1966," *Wired*, September 27, 2013, https://www.wired.com/2013/09/tech-time-warp-shakey-robot (accessed November 20, 2020).

11. Planning sequences of actions is found in primates and also in corvids. See Elizabeth Pennisi, "Ravens—like Humans and Apes—Can Plan for the Future," *Science*, July 13, 2017, https://www.sciencemag.org/news/2017/07/ravens-humans-and-apes-can-plan-future (accessed November 20, 2020).

12. See note 4 in chap. 4 above.

13. For the CMU CoBots, see M. M. Veloso, J. Biswas, B. Coltin, and S. Rosenthal, "CoBots: Robust Symbiotic Autonomous Mobile Service Robots," *IJCAI*, July 2015, 4423, https://pdfs.semanticscholar.org/468d/43734488e0e29c3c11f2c15d9b1fb6f1adc4.pdf (accessed November 20, 2020).

14. E.g., Peter Jackson, *Introduction to Expert Systems*, 3rd ed. (Addison-Wesley, 1998).

15. MYCIN was an early expert system dealing with bacterial blood infections; its approach was later developed into a comprehensive system called INTERNIST and then CADUCEUS. G. Banks, "Artificial Intelligence in Medical Diagnosis: The INTERNIST/CADUCEUS Approach," *Critical Reviews in Medical Informatics* 1, no. 1 (1986): 23–54. XCON was an early expert system for configuration. Used by computer companies to assemble minicomputer components, the approach is now widely applied in modular home building, manufacturing, many online custom configuration systems, and large business environments such as SAP. See, e.g., A. Felfernig, L. Hotz, C. Bagley, and J. Tiihonen, *Knowledge-Based Configuration: From Research to Business Cases* (Morgan Kaufmann, 2014). Like much practical AI, it is no longer thought of as AI.

16. Industrial robots must move accurately in their specific engineered environment and are often programmed with a *teach pendant*. This handheld device allows an engineer to move a robot arm from point to point and record the movements for later use.

17. A large number of small, cheap robots are available from the long-established Lego Mindstorms, ranging from tiny self-build robot arms to innovative shapes like a rolling spherical robot.

18. The saying is widely attributed to Einstein but without evidence that he in fact said this.

19. Olivia Solon, "Roomba Creator Responds to Reports of 'Poopocalypse': 'We See This a Lot,'" *The Guardian*, August 15, 2016, https://www.theguardian.com/technology/2016/aug/15/roomba-robot-vacuum-poopocalypse-facebook-post. For those with Facebook access, a graphic account from the victim here: https://www.facebook.com/jesse.newton.37/posts/776177951574 (accessed November 20, 2020).

CHAPTER 8

1. Karl Sims, "Evolving Virtual Creatures," in *Proceedings of the 21st Annual Conference on Computer Graphics and Interactive Techniques* (ACM, July 1994), 15–22, https://www.karlsims.com/papers/siggraph94.pdf. For Blockies in action, see https://www.youtube.com/watch?v=JBgG_VSP7f8 (accessed November 20, 2020).

2. For a useful tutorial on genetic algorithms by one of the pioneers of the field, John Holland, see http://www2.econ.iastate.edu/tesfatsi/holland.gaintro.htm (accessed November 20, 2020).

3. J. B. Pollack and H. Lipson, "The GOLEM Project: Evolving Hardware Bodies and Brains," in *Proceedings: The Second NASA/DoD Workshop on Evolvable Hardware* (IEEE, July 2000), 37–42, https://www.researchgate.net/publication/3864695_The_GOLEM_project_evolving_hardware_bodies_and_brains (accessed November 20, 2020).

4. This work was part of the active field called *evolutionary robotics*. See Agoston E. Eiben, "Grand Challenges for Evolutionary Robotics," *Frontiers in Robotics and AI*, June 30, 2014, https://www.frontiersin.org/articles/10.3389/frobt.2014.00004/full#B14 (accessed November 20, 2020).

5. For a survey of reinforcement learning in robots, see J. Kober, J. A. Bagnell, and J. Peters, "Reinforcement Learning in Robotics: A Survey," *International Journal of Robotics Research* 32, no. 11 (2013): 1238–1274, http://www.cs.cmu.edu/~jeanoh/16-785/papers/kober-ijrr2013-rl-in-robotics-survey.pdf (accessed November 20, 2020).

6. An interesting blog post by Alex Kendall, a researcher at Cambridge, discusses the specific issues of applying reinforcement learning to real-world robots: "Now Is the Time for Reinforcement Learning on Real Robots," updated April 23, 2019, https://alexgkendall.com

/reinforcement_learning/now_is_the_time_for_reinforcement_learning _on_real_robots (accessed November 20, 2020).

7. A blog article by Alex Irpan, a researcher into reinforcement learning, points out that it took eighteen million examples for one RL system to learn good moves in an Atari game that a human picks up in tens of minutes: "Deep Reinforcement Learning Doesn't Work Yet," February 14, 2018, https://www.alexirpan.com/2018/02/14/rl -hard.html (accessed November 20, 2020).

8. For the Fred and Ginger robots, see D. P. Barnes, R. A. Ghanea-Hercock, R. Aylett, and A. M. Coddington, "Many Hands Make Light Work? An Investigation into Behaviorally Controlled Co-operant Autonomous Mobile Robots," *Agents*, February 1997, 413–420, http:// citeseerx.ist.psu.edu/viewdoc/download?doi=10.1.1.53.3565&rep=rep1 &type=pdf (accessed November 20, 2020).

9. Kendall, "Now Is the Time."

10. Irpan, "Deep Reinforcement Learning."

11. A thought experiment of Nick Bostrom, a Swedish philosopher, who discusses the necessity of ethical constraints on robot behavior. N. Bostrom, "Ethical Issues in Advanced Artificial Intelligence," in *Science Fiction and Philosophy: From Time Travel to Superintelligence* (London: Wiley-Blackwell, 2009), 277–284.

12. Marvin Minsky and Seymour Papert, *Perceptrons: An Introduction to Computational Geometry* (Cambridge, MA: MIT Press, 1969). Perceptrons are credited with killing the initial enthusiasm for ANNs.

13. The backpropagation algorithm was conceived back in the 1960s but never published; most researchers learned about it via D. E. Rumelhart, G. E. Hinton, and R. J. Williams, "Learning Internal Representations by Error Propagation," in *Parallel Distributed Processing,* ed. D. E. Rumelhart and J. L. McClelland (Cambridge, MA: MIT Press, 1986).

14. The philosopher Hubert Dreyfus was a severe critic of AI hype, arguing that the whole approach was fundamentally mistaken. His last book on the topic was *What Computers Still Can't Do: A Critique of Artificial Reason* (Cambridge, MA: MIT Press, 1992). He died in 2017, but see an article by Gerben Wierda, arguing that this is all still true, in the online magazine *InfoWorld,* March 21, 2018, https://www.infoworld .com/article/3263755/something-is-still-rotten-in-the-kingdom-of -artificial-intelligence.html (accessed November 20, 2020).

15. For the mislabeling of Google photos, see James Vincent, "Google 'Fixed' Its Racist Algorithm by Removing Gorillas from Its Image-Labeling Tech," *The Verge*, January 12, 2018, https://www.theverge.com/2018/1/12/16882408/google-racist-gorillas-photo-recognition-algorithm-ai (accessed November 20, 2020). It appears that the basic error is yet to be fixed.

16. For the wolves and dogs example, see Gary Marcus, "In Defense of Skepticism about Deep Learning," Medium.com, January 14, 2018, https://medium.com/@GaryMarcus/in-defense-of-skepticism-about-deep-learning-6e8bfd5ae0f1 (accessed November 20, 2020).

17. For OpenAI's own popular account of the Rubik's Cube robot, see https://openai.com/blog/solving-rubiks-cube (accessed November 20, 2020). For the complete research paper, see I. Akkaya, M. Andrychowicz, M. Chociej, M. Litwin, B. McGrew, A. Petron, A. Paino, et al., "Solving Rubik's Cube with a Robot Hand," October 17, 2019, https://arxiv.org/pdf/1910.07113.pdf (accessed November 20, 2020).

18. Many online reviews of the work with the Rubik's Cube point out both its successes and its limitations; see, e.g., Will Knight, "Why Solving a Rubik's Cube Does Not Signal Robot Supremacy," *Wired*, October 16, 2019, https://www.wired.com/story/why-solving-rubiks-cube-not-signal-robot-supremacy (accessed November 20, 2020).

19. For an article on an MIT robot learning to play Jenga, see Matt Simon, "A Robot Teaches Itself to Play Jenga. But This Is No Game," *Wired*, January 30, 2019, https://www.wired.com/story/a-robot-teaches-itself-to-play-jenga (accessed November 20, 2020). For more technical detail, see N. Fazeli, M. Oller, J. Wu, Z. Wu, J. B. Tenenbaum, and A. Rodriguez, "See, Feel, Act: Hierarchical Learning for Complex Manipulation Skills with Multisensory Fusion," *Science Robotics* 4, no. 26 (2019): 3123, https://jiajunwu.com/papers/jenga_scirobot.pdf (accessed November 20, 2020).

20. For more about the iCub robot, see http://www.icub.org (accessed November 20, 2020).

21. For work on crawling, see L. Righetti and A. J. Ijspeert, "Design Methodologies for Central Pattern Generators: An Application to Crawling Humanoids," in *Proceedings of Robotics: Science and Systems* (2006), 191–198, http://robotcub.org/misc/papers/06_Righetti_Ijspeert_RSS.pdf (accessed November 20, 2020).

22. For a comprehensive survey of developmental robotics, see M. Asada, K. Hosoda, Y. Kuniyoshi, H. Ishiguro, T. Inui, Y. Yoshikawa, M. Ogino, and C. Yoshida, "Cognitive Developmental Robotics: A Survey," *IEEE Transactions on Autonomous Mental Development* 1, no. 1 (2009): 12–34, https://www.cs.tufts.edu/comp/150DR/readings/week1/Asada09g.pdf (accessed November 20, 2020).

23. For an overview of Grey Walter and his work, see O. Holland, "The First Biologically Inspired Robots," *Robotica* 21, no. 4 (2003): 351–363, https://pdfs.semanticscholar.org/d992/8f7c0f91bde0e9364e8dc997749e0c5f10b4.pdf (accessed November 20, 2020).

24. For an overview of the field, see P. van der Smagt, M. A. Arbib, and G. Metta, "Neurorobotics: From Vision to Action," in *Springer Handbook of Robotics* (Springer, 2016), 2069–2094, https://mediatum.ub.tum.de/doc/1289380/file.pdf (accessed November 20, 2020).

25. For further information on the project that applies neurorobotics to Parkinson's disease, see http://www.macs.hw.ac.uk/neuro4pd (accessed November 20, 2020).

CHAPTER 9

1. For an article by the late Martin Gardner, a well-known columnist on mathematical issues, see "Mathematical Games: The Fantastic Combinations of John Conway's New Solitaire Game 'Life,'" *Scientific American* 223 (October 1970): 120–123, https://web.stanford.edu/class/sts145/Library/life.pdf (accessed November 20, 2020). There are many implementations online; see Google Play for Android devices and the Apple App Store for IOS.

2. Emergence is one of many ideas in complexity science, which declares itself as an alternative approach to Cartesian reductionism. For a good introduction to emergence and many other aspects of this field, see Melanie M. Mitchell, *Complexity: A Guided Tour* (Oxford University Press, 2009).

3. Argued in the well-known popular science book by D. R. Hofstadter, *Gödel, Escher, Bach: An Eternal Golden Braid* (New York: Vintage, 1979).

4. The biological ideas underpinning what is now called swarm robotics are discussed in S. Garnier, J. Gautrais, and G. Theraulaz, "The Biological

Principles of Swarm Intelligence," *Swarm Intelligence* 1, no. 1 (2007): 3–31, https://www.researchgate.net/profile/Simon_Garnier/publication/220058931_The_biological_principles_of_swarm_intelligence/links/09e41507701a2e0675000000.pdf (accessed November 20, 2020).

5. For desire paths, see Kurt Kohlstedt, "Least Resistance: How Desire Paths Can Lead to Better Design," *99% Invisible*, January 25, 2016, https://99percentinvisible.org/article/least-resistance-desire-paths-can-lead-better-design (accessed November 20, 2020). He suggests that some universities and other public bodies in the United States have used desire paths as a guide to laying more permanent paths.

6. Garnier, Gautrais, and Theraulaz, "Biological Principles of Swarm Intelligence."

7. Ant algorithms are well used outside robotics in scheduling, data telecommunication routing, image processing, and other fields. Jean-Louis Deneubourg and collaborators carried out a set of experiments with ants in the late 1980s and early 1990s, and Dorigo and others generalized this as a set of algorithms in the mid-1990s. M. Dorigo, V. Maniezzo, and A. Colorni, "The Ant System: Optimization by a Colony of Cooperating Agents," *IEEE Transactions on Systems, Man, and Cybernetics, Part B: Cybernetics* 26, no. 1 (1996): 29–41, http://www.cs.unibo.it/babaoglu/courses/cas05-06/tutorials/Ant_Colony_Optimization.pdf (accessed November 20, 2020).

8. For example, researchers at Harvard University have developed the Kilobot and experimented with 1,024 of these robots. "A Swarm of One Thousand Robots," *IEEE Spectrum*, video posted to YouTube, August 14, 2014, https://www.youtube.com/watch?v=G1t4M2XnIhI (accessed November 20, 2020). Each has three behaviors, moves via vibration, and communicates with reflected infrared.

9. Box pushing has been a standard experimental task almost from the start of swarm robotics research; see, e.g., C. R. Kube and H. Zhang, "Collective Robotics: From Social Insects to Robots," *Adaptive Behavior* 2, no. 2 (1993): 189–218, http://biorobotics.ri.cmu.edu/papers/sbp_papers/integrated1/kube_collective_robotics.pdf (accessed November 20, 2020).

10. "Cancer-Fighting Nanorobots Programmed to Seek and Destroy Tumors," *Science Daily*, February 12, 2018, https://www.sciencedaily.com/releases/2018/02/180212112000.htm (accessed November 20, 2020).

11. The paper introducing this idea is Craig Reynolds, "Flocks, Herds and Schools: A Distributed Behavioral Model," in *SIGGRAPH '87: Proceedings of the 14th Annual Conference on Computer Graphics and Interactive Techniques* (New York: Association for Computing Machinery, 1987), 25–34.

12. How to carry out swarm engineering is the topic of A. F. T. Winfield, C. J. Harper, and J. Nembrini, "Towards Dependable Swarms and a New Discipline of Swarm Engineering," in *Swarm Robotics Workshop: State-of-the-Art Survey*, vol. 3342, ed. Erol Şahin and William Spears (Berlin: Springer, 2005), 126–142, http://citeseerx.ist.psu.edu/viewdoc/download?doi=10.1.1.182.8033&rep=rep1&type=pdf (accessed November 20, 2020).

13. For the aims of the initiative, see H. Kitano, M. Asada, Y. Kuniyoshi, I. Noda, E. Osawa, and H. Matsubara, "RoboCup: A Challenge Problem for AI," *AI Magazine* 18, no. 1 (1997): 73–73, http://citeseerx.ist.psu.edu/viewdoc/download?doi=10.1.1.662.6314&rep=rep1&type=pdf (accessed November 20, 2020). RoboCup itself is at https://www.robocup.org (accessed November 20, 2020). It is a very open organization, and each different league has its own web presence.

14. For an article about the small robot subleague, see A. Weitzenfeld, J. Biswas, M. Akar, and K. Sukvichai, "RoboCup Small-Size League: Past, Present and Future," in *Robot Soccer World Cup* (Springer, July 2014), 611–623, https://link.springer.com/chapter/10.1007/978-3-319-18615-3_50 (accessed November 20, 2020).

15. For video of the Small Size League in action in 2015, see https://www.youtube.com/watch?v=HhikJB24m7M (accessed November 20, 2020).

16. One of the most successful teams over the years is Carnegie Mellon University's CMDragons. For an article reviewing their experience up to 2013, see J. Biswas, J. P. Mendoza, D. Zhu, B. Choi, S. Klee, and M. Veloso, "Opponent-Driven Planning and Execution for Pass, Attack, and Defense in a Multi-robot Soccer Team," in *Proceedings 2014 International Conference on Autonomous Agents and Multi-agent Systems* (IFAAM, May 2014), 493–500, http://aamas.csc.liv.ac.uk/Proceedings/aamas2014/aamas/p493.pdf (accessed November 20, 2020).

17. The Standard Platform League uses the Soft Bank Nao robots, humanoids about 0.6 meters tall. For highlights of them in action in 2018, see https://www.youtube.com/watch?v=pmFKoKtRW6s&vl=en (accessed November 20, 2020).

18. For a useful article about the RoboCup Robot Rescue competition, see H. L. Akin, N. Ito, A. Jacoff, A. Kleiner, J. Pellenz, and A. Visser, "RoboCup Rescue Robot and Simulation Leagues," *AI Magazine* 34, no. 1 (2013): 78–86, https://pure.uva.nl/ws/files/1996468/150144_Robo-Cup_Rescue_Robot_and_Simulation_Leagues.pdf (accessed November 20, 2020).

19. The German company Telerob, https://www.telerob.com/en/news-media (accessed November 20, 2020).

20. The Center for Robot-Assisted Search and Rescue (CRASAR) is at http://crasar.org. For other interventions, see Daniel Faggella, "11 Robotic Applications for Search and Rescue," *Huffington Post*, November 23, 2017, https://www.huffpost.com/entry/11-robotic-applications-for-search-and-rescue_b_5a173c9ae4b0bf1467a845c4 (accessed November 20, 2020).

21. The classic discussion of levels of autonomy in teleoperation is T. B. Sheridan and W. L. Verplank, "Human and Computer Control of Undersea Teleoperators," Tech. Rep., DTIC Document, 1978, https://apps.dtic.mil/dtic/tr/fulltext/u2/a057655.pdf (accessed November 20, 2020).

22. For a discussion of these issues applied to autonomous underwater vehicles, see H. Hastie, K. Lohan, M. Chantler, D. A. Robb, S. Ramamoorthy, R. Petrick, S. Vijayakumar, and D. Lane, "The Orca Hub: Explainable Offshore Robotics through Intelligent Interfaces," March 6, 2018, arXiv preprint arXiv:1803.02100, https://arxiv.org/pdf/1803.02100.pdf (accessed November 20, 2020).

23. Daniel Dennett, a philosopher writing on consciousness and other topics, describes this as "the intentional stance."

24. A. Malik, F. Giones, and T. Schweisfurth, "Meet the Cobots: The Robots Who Will Be Your Colleagues, Not Your Replacements," *The Conversation*, October 29, 2019, http://theconversation.com/meet-the-cobots-the-robots-who-will-be-your-colleagues-not-your-replacements-125189 (accessed November 20, 2020).

CHAPTER 10

1. For the deployment of CIMON, see Mike Wehner, "The International Space Station Is Getting a Floating AI Assistant, and It Sure Looks Familiar," *BGR*, March 2, 2018, https://bgr.com/2018/03/02/cimon-iss-ai-space-station-nasa (accessed November 20, 2020).

2. For a video of CIMON's odd behavior, see Mike Wehner, "ESA's Adorable Space Station AI Had an Emotional Meltdown in His Debut," *BGR*, December 3, 2018, https://bgr.com/2018/12/03/cimon-ai-emotional-meltdown-iss (accessed November 20, 2020).

3. Defined in the American Psychological Association Dictionary of Psychology at https://dictionary.apa.org/affect as "any experience of feeling or emotion, ranging from suffering to elation, from the simplest to the most complex sensations of feeling, and from the most normal to the most pathological emotional reactions."

4. The neurophysiologist Antonio Damasio, in *Descartes' Error: Emotion, Reason, and the Human Brain* (Putnam, 1994), argued influentially against Cartesian mind-body dualism and contended that rationality required emotion. His argument is summarized at https://pdfs.semanticscholar.org/29de/be35fb6cbe3cdede9a6f0e993681874bc8ec.pdf (accessed November 20, 2020).

5. For a recent popular science account of emotion, see L. F. Barrett, *How Emotions Are Made: The Secret Life of the Brain* (Houghton Mifflin Harcourt, 2017).

6. The grouping of emotions along dimensions later known as *arousal* and *valence* was the work of James Russell in 1980. Cards with emotion names on them were given to a large number of people, who were asked to sort them into groups. In the original classification, anger and fear appeared next to each other, when for most of us these emotions feel very different. A third dimension, called *dominance*, was added to differentiate between them. The whole thing is discussed in A. Mehrabian, "Pleasure-Arousal-Dominance: A General Framework for Describing and Measuring Individual Differences in Temperament," *Current Psychology* 14, no. 4 (1996): 261–292.

7. For work on developing a curious robot, see P. Y. Oudeyer, F. Kaplan, V. V. Hafner, and A. Whyte, "The Playground Experiment: Task-Independent Development of a Curious Robot," 2004, https://core.ac.uk/download/pdf/22873818.pdf (accessed November 20, 2020).

8. For a good summary of cognitive appraisal, see I. J. Rosemanand and C. A. Smith, "Appraisal Theory: Overview, Assumptions, Varieties, Controversies," in *Appraisal Processes in Emotion: Theory, Methods, Research*, ed. K. R. Scherer, A. Schorr, and T. Johnstone, Series in Affective Science (Oxford University Press, 2001), 3–19.

9. The most influential approach to cognitive appraisal for computational modeling was that of Ortony, Clore, and Collins in 1984, so often applied that it is abbreviated to OCC. A. Ortony, G. L. Clore, and A. Collins, *The Cognitive Structure of Emotions* (Cambridge University Press, 1990).

10. Coping behavior is discussed by Richard Lazarus in his 1980s work. R. S. Lazarus, "Coping Theory and Research: Past, Present, and Future," in *Fifty Years of the Research and Theory of R. S. Lazarus: An Analysis of Historical and Perennial Issues* (Mayweh, NJ: Laurence Erlbaum Associates, 1998), 366–388, http://emotionalcompetency.com/papers/coping%20research.pdf (accessed November 20, 2020).

11. For the iCat Chess Companion, see I. Leite, A. Pereira, C. Martinho, A. Paiva, and G. Castellano, "Towards an Empathic Chess Companion," *Autonomous Agents and Multiagent Systems (AAMAS 2009)*, 2009, http://www.inesc-id.pt/ficheiros/publicacoes/6190.pdf (accessed November 20, 2020).

12. Facial action units are widely used in psychology to annotate videos, a frame at a time, of people interacting with each other, forming a Facial Action Coding System (FACS).

13. As with many emotion theories in psychology, not all psychologists agree. Richard Ekman, who originated the idea, reported that all these expressions were recognized—from photographs—in cultures as far apart as the United States and New Guinea. R. Ekman, *What the Face Reveals: Basic and Applied Studies of Spontaneous Expression Using the Facial Action Coding System (FACS)* (Oxford University Press, 1997). Opponents argue that some of them were regularly confused—especially surprise and disgust—and that the context in which an expression is set has a substantial impact on how people classify it.

14. Ken Perlin, a well-known researcher into animation, has Polly World—blocks with personality—among many other things at his web pages at https://mrl.nyu.edu/~perlin.

15. The commercial robots Pepper and Nao from SoftBank do not have any facial movement, but around their eyes they do have colored LEDs that can be used to communicate affective state.

16. For the empathic tutor, see M. Obaid, R. Aylett, W. Barendregt, C. Basedow, L. J. Corrigan, L. Hall, A. Jones, A. Kappas, D. Küster, A. Paiva, and F. Papadopoulos, "Endowing a Robotic Tutor with Empathic

Qualities: Design and Pilot Evaluation," *International Journal of Humanoid Robotics* 15, no. 6 (2018): 1850025, https://sure.sunderland.ac.uk/id/eprint/10074 (accessed November 20, 2020).

17. For a survey of work on multimodal emotion recognition, see T. Baltrušaitis, C. Ahuja, and L. P. Morency, "Multimodal Machine Learning: A Survey and Taxonomy," *IEEE Transactions on Pattern Analysis and Machine Intelligence* 41, no. 2 (2018): 423–443, https://arxiv.org/pdf/1705.09406.pdf (accessed November 20, 2020).

18. For an IBM in-house account of Watson, see D. A. Ferrucci, introduction to "This Is Watson," *IBM Journal of Research and Development* 56, nos. 3–4 (2012): 1–15, http://brenocon.com/watson_special_issue/01%20Intro.pdf (accessed November 20, 2020).

CHAPTER 11

1. Masahiro Fujita, who conceived the Aibo back in the early 1990s, wrote about his approach in "AIBO: Toward the Era of Digital Creatures," *International Journal of Robotics Research* 20, no. 10 (2001): 781–794, https://journals.sagepub.com/doi/pdf/10.1177/02783640122068092 (accessed November 20, 2020).

2. A team at the University of Washington analyzed 6,438 postings on Aibo online forums to examine how owners felt about their Aibos. B. Friedman, P. H. Kahn Jr., and J. Hagman, "Hardware Companions? What Online AIBO Discussion Forums Reveal about the Human-Robotic Relationship," in *Proc SIGCHI Conference Human Factors in Computing Systems* (ACM, April 2003), 273–280, https://www.vsdesign.org/publications/pdf/friedman03hardwarecompanions.pdf (accessed November 20, 2020).

3. The *Japan Times* reported encouraging initial sales; see "Sales of Sony's New Aibo Robot Dog Off to Solid Start," May 7, 2018, https://www.japantimes.co.jp/news/2018/05/07/business/tech/sales-sonys-new-aibo-robot-dog-off-solid-start/#.XiIhrS10chs (accessed November 20, 2020).

4. For example, a 2013 study in ten US nursing homes showed a continuing effect. S. Šabanović, C. C. Bennett, W. L. Chang, and L. Huber, "PARO Robot Affects Diverse Interaction Modalities in Group Sensory Therapy for Older Adults with Dementia," in *2013 IEEE 13th International Conference on Rehabilitation Robotics (ICORR)* (IEEE, June

2013), 1–6, http://homes.sice.indiana.edu/selmas/Sabanovic-ICORR13.pdf (accessed November 20, 2020).

5. See the discussion in I. Leite, C. Martinho, A. Pereira, and A. Paiva, "As Time Goes By: Long-Term Evaluation of Social Presence in Robotic Companions," in *RO-MAN 2009—18th IEEE International Symposium on Robot and Human Interactive Communication* (IEEE, 2009), 669–674, http://www.inesc-id.pt/ficheiros/publicacoes/6182.pdf (accessed November 20, 2020).

6. Bridget Carey, "My Week with Aibo: What It's Like to Live with Sony's Robot Dog," *C/NET*, November 28, 2018, https://www.cnet.com/news/my-week-with-aibo-what-its-like-to-live-with-sonys-robot-dog (accessed November 20, 2020).

7. You can see the Lindsey robot at https://www.youtube.com/watch?v=x6rA5E_Belk (accessed November 20, 2020) and read about it in F. Del Duchetto, P. Baxter, and M. Hanheide, "Lindsey the Tour Guide: Robot-Usage Patterns in a Museum Long-Term Deployment," in *2019 28th IEEE International Conference on Robot and Human Interactive Communication (RO-MAN)* (IEEE, October 2019), 1–8, http://eprints.lincoln.ac.uk/37348/1/RO_MAN_Lindsey%20%281%29.pdf (accessed November 20, 2020).

8. For children bullying a robot in a shopping mall, see D. Brscić, H. Kidokoro, Y. Suehiro, and T. Kanda, "Escaping from Children's Abuse of Social Robots," in *Proceedings 10th Annual ACM/IEEE International Conference on Human-Robot Interaction* (ACM, March 2015), 59–66, http://citeseerx.ist.psu.edu/viewdoc/download?doi=10.1.1.714.7880&rep=rep1&type=pdf (accessed November 20, 2020).

9. For the hitchBOT, see https://web.archive.org/web/20140809025115/http://www.hitchbot.me/wp-content/media/hB_MediaKit_Summer2014.pdf and for its destruction in Philadelphia, see Dominique Mosbergen, "Good Job, America. You Killed hitchBOT," *HuffPost*, August 3, 2015, https://www.huffingtonpost.co.uk/entry/hitchbot-destroyed-philadelphia_n_55bf24cde4b0b23e3ce32a67 (accessed November 20, 2020).

10. For the fire evacuation experiment, see P. Robinette, W. Li, R. Allen, A. M. Howard, and A. R. Wagner, "Overtrust of Robots in Emergency Evacuation Scenarios," in *11th ACM/IEEE International Conference on Human-Robot Interaction* (IEEE Press, March 2016), 101–108,

https://sites.psu.edu/real/files/2016/08/Robinette-HRI-2016-1wswob0.pdf (accessed November 20, 2020).

11. For Bert, the omelet-making robot, see A. Hamacher, N. Bianchi-Berthouze, A. G. Pipe, and K. Eder, "Believing in BERT: Using Expressive Communication to Enhance Trust and Counteract Operational Error in Physical Human-Robot Interaction," in *25th IEEE International Symposium on Robot and Human Interactive Communication (RO-MAN)* (IEEE, August 2016), 493–500, https://arxiv.org/ftp/arxiv/papers/1605/1605.08817.pdf (accessed November 20, 2020).

12. For experiments involving a robot approaching someone sitting at a table with an object the person had asked for, see K. Dautenhahn, M. Walters, S. Woods, K. L. Koay, C. L. Nehaniv, A. Sisbot, R. Alami, and T. Siméon, "How May I Serve You? A Robot Companion Approaching a Seated Person in a Helping Context," in *Proceedings 1st ACM SIGCHI/SIGART Conference on Human-Robot Interaction* (ACM, 2006), 172–179, https://hal.laas.fr/hal-01979221/document (accessed November 20, 2020).

13. A table of comfortable distances is provided in M. L. Walters, K. Dautenhahn, R. Te Boekhorst, K. L. Koay, D. S. Syrdal, and C. L. Nehaniv, "An Empirical Framework for Human-Robot Proxemics," *Proceedings of New Frontiers in Human-Robot Interaction*, 2009, https://uhra.herts.ac.uk/bitstream/handle/2299/9670/903515.pdf (accessed November 20, 2020).

14. For work on how fast a robot arm should move for human comfort, see M. K. Pan, E. Knoop, M. Bächer, and G. Niemeyer, "Fast Handovers with a Robot Character: Small Sensorimotor Delays Improve Perceived Qualities," *IEEE/RSJ International Conference on Intelligent Robots and Systems (IROS)*, 2019, https://la.disneyresearch.com/publication/fast-handovers-with-a-robot-character-small-sensorimotor-delays-improve-perceived-qualities (accessed November 20, 2020).

15. A short article by Rozita Dara discusses some of the privacy concerns with current digital assistants. See "The Dark Side of Alexa, Siri and Other Personal Digital Assistants," *The Conversation*, December 15, 2019, http://theconversation.com/the-dark-side-of-alexa-siri-and-other-personal-digital-assistants-126277 (accessed November 20, 2020).

16. For discussion of how to build a memory for a long-lived robot companion, see M. Y. Lim, R. Aylett, W. C. Ho, S. Enz, and P. Vargas,

"A Socially-Aware Memory for Companion Agents," in *International Workshop on Intelligent Virtual Agents* (Springer, September 2009), 20–26, https://www.researchgate.net/publication/221588267_A_Socially-Aware_Memory_for_Companion_Agents (accessed November 20, 2020).

17. Socially assistive robotics has become a well-defined research field. For an overview, see M. J. Matarić and B. Scassellati, "Socially Assistive Robotics," in *Springer Handbook of Robotics* (Springer, 2016), 1973–1994, https://robotics.usc.edu/publications/media/uploads/pubs/pubdb_1045_daf14ca731584017a19ca751e7459f4a.pdf (accessed November 20, 2020).

CHAPTER 12

1. There was a great deal of publicity at the time, none of it good for Microsoft. For a more detached discussion, see G. Neff and P. Nagy, "Automation, Algorithms, and Politics Talking to Bots: Symbiotic Agency and the Case of Tay," *International Journal of Communication* 10, no. 17 (2016): 4915–4931, https://ijoc.org/index.php/ijoc/article/viewFile/6277/1804 (accessed November 20, 2020).

2. Joseph Weizenbaum was the creator of Eliza: this obituary in 2008 gives an overview of how he felt about his creation: https://www.independent.co.uk/news/obituaries/professor-joseph-weizenbaum-creator-of-the-eliza-program-797162.html (accessed November 20, 2020).

3. For how users feel about interacting with chatbots, see P. B. Brandtzaeg and A. Følstad, "Why People Use Chatbots," in *International Conference on Internet Science* (Springer, 2017), 377–392, https://sintef.brage.unit.no/sintef-xmlui/bitstream/handle/11250/2468333/Brandtzaeg_Folstad_why+people+use+chatbots_authors+version.pdf (accessed November 20, 2020).

4. For the problems of asking voice assistants for medical advice, see T. W. Bickmore, H. Trinh, S. Olafsson, T. K. O'Leary, R. Asadi, N. M. Rickles, and R. Cruz, "Patient and Consumer Safety Risks When Using Conversational Assistants for Medical Information: An Observational Study of Siri, Alexa, and Google Assistant," *Journal of Medical Internet Research* 20, no. 9 (2018): e11510, https://www.jmir.org/2018/9/e11510 (accessed November 20, 2020).

5. For the 2018 Alexa Challenge, the competitors, and the results, see https://developer.amazon.com/alexaprize/challenges/past-challenges/2018 (accessed November 20, 2020).

6. For a discussion of abusive interaction with conversational agents and thoughts on how to deal with it, see A. C. Curry and V. Rieser, "A Crowd-Based Evaluation of Abuse Response Strategies in Conversational Agents," *Proceedings of the 20th Annual SIGdial Meeting on Discourse and Dialogue,* September 2019, 361–366, https://www.sigdial.org/files/workshops/conference20/proceedings/cdrom/pdf/W19-5942.pdf (accessed November 20, 2020).

7. Researchers in the United States are building mobile robots for military use that could follow spoken instructions and report back on their task. However, they need to hold explicit knowledge about the environment they must operate in. David Hambling, "The US Army Is Creating Robots That Can Follow Orders," *MIT Technology Review,* November 6, 2019, https://www.technologyreview.com/s/614686/the-us-army-is-creating-robots-that-can-follow-ordersand-ask-if-they-dont-understand (accessed November 20, 2020).

8. Google researchers report on their work: C. C. Chiu, T. N. Sainath, Y. Wu, R. Prabhavalkar, P. Nguyen, Z. Chen, A. Kannan, R. J. Weiss, K. Rao, E. Gonina, and N. Jaitly, "State-of-the-Art Speech Recognition with Sequence-to-Sequence Models," in *2018 IEEE International Conference on Acoustics, Speech and Signal Processing (ICASSP)* (IEEE, April 2018), 4774–4778, https://arxiv.org/pdf/1712.01769.pdf (accessed November 20, 2020).

9. Recent work tries to add expression to synthetic voices; see C. G. Buchanan, M. P. Aylett, and D. A. Braude, "Adding Personality to Neutral Speech Synthesis Voices," in *International Conference on Speech and Computer* (Springer, September 2018), 49–57, https://www.researchgate.net/profile/Christopher_Buchanan3/publication/327845291_Adding_Personality_to_Neutral_Speech_Synthesis_Voices/links/5dc0a3e9a6fdcc21280478ca/Adding-Personality-to-Neutral-Speech-Synthesis-Voices.pdf (accessed November 20, 2020).

10. For an experiment comparing two robots with different voices, see H. Hastie, K. Lohan, A. Deshmukh, F. Broz, and R. Aylett, "The Interaction between Voice and Appearance in the Embodiment of a Robot Tutor," in *International Conference on Social Robotics* (Springer,

November 2017), 64–74, https://pureapps2.hw.ac.uk/ws/portalfiles/portal/15871745/ICSR2017_final.pdf (accessed November 20, 2020).

11. This is the standard interpretation of what Turing was saying; he discussed the subject in more than one paper, so some feel his views changed over time and were not quite as this version of the test assumes.

12. Wikipedia is a good source for details of the Loebner Prize and contestants over the years; see https://en.wikipedia.org/wiki/Loebner_Prize (accessed November 20, 2020).

13. For a summary of the Chinese room and responses to it, see D. Cole, "The Chinese Room Argument," *Stanford Encyclopedia of Philosophy*, March 19, 2004, revised April 9, 2014, https://plato.stanford.edu/entries/chinese-room (accessed November 20, 2020).

14. An example cited by the Russian psychologist Lev Vygotsky, who worked on theories of language development. L. S. Vygotsky, *Thinking and Speech* (1934), chap. 7, 270–271, https://www.marxists.org/archive/vygotsky/works/words/Thinking-and-Speech.pdf (accessed November 20, 2020).

15. For the role of language games in grounding language for robots, see L. Steels, "Language Games for Autonomous Robots," *IEEE Intelligent Systems* 16, no. 5 (2001): 16–22, https://digital.csic.es/bitstream/10261/128135/1/Language%20games.pdf (accessed November 20, 2020).

16. Steels, "Language Games for Autonomous Robots."

CHAPTER 13

1. You can see this interview of April 26, 2017, at https://www.youtube.com/watch?v=Bg_tJvCA8zw (accessed November 20, 2020). Try watching it twice, the second time with the sound off, to assess the body language involved.

2. Professor Noel Sharkey, a senior robotics researcher in the United Kingdom, points out that Sophia is "a show robot"—a marketing device—and regrets what he perceives as intentional deception about its capabilities in "Mama Mia It's Sophia: A Show Robot or a Dangerous Platform to Mislead?" *Forbes*, November 17, 2018, https://www.forbes.com/sites/noelsharkey/2018/11/17/mama-mia-its-sophia-a-show-robot-or-dangerous-platform-to-mislead.

3. Concerns about automated decision systems, especially those based on an opaque machine learning approach, are growing. In the Netherlands, an AI-based system introduced in 2014 to identify potential welfare fraudsters was recently ruled illegal. The court decided that the system violated EU law on privacy and human rights. Don Jacobson, "Dutch Anti-fraud System Violates Human Rights, Court Rules," UPI, February 5, 2020, https://www.upi.com/Top_News/World-News/2020/02/05/Dutch-anti-fraud-system-violates-human-rights-court-rules/6051580914081 (accessed November 20, 2020).

4. For the IEEE work, see https://standards.ieee.org/industry-connections/ec/autonomous-systems.html (accessed November 20, 2020).

5. The European Union (EU) has produced a set of Ethics Guidelines for Trustworthy AI. See https://ec.europa.eu/digital-single-market/en/news/ethics-guidelines-trustworthy-ai (accessed November 20, 2020).

6. For the work of specialists in the United Kingdom on ethical principles for robotics, see https://epsrc.ukri.org/research/ourportfolio/themes/engineering/activities/principlesofrobotics (accessed November 20, 2020).

7. For a critical account of the robot "giving evidence," see James Vincent, "The UK Invited a Robot to 'Give Evidence' in Parliament for Attention," *The Verge*, October 12, 2018, https://www.theverge.com/2018/10/12/17967752/uk-parliament-pepper-robot-invited-evidence-select-committee (accessed November 20, 2020).

8. The International Committee for Robot Arms Control, with discussion materials on LARs, https://www.icrac.net (accessed November 20, 2020).

9. The Campaign Against Killer Robots, https://www.stopkillerrobots.org (accessed November 20, 2020).

10. Peter Fussey and Daragh Murray, "Independent Report on the London Metropolitan Police Service's Trial of Live Facial Recognition Technology," project report, University of Essex Human Rights Centre, 2019. The report is not publicly available, but its contents are reported in Rachel England, "UK Police's Facial Recognition System Has an 81 Percent Error Rate," *Engadget*, April 7, 2019, https://www.engadget.com/2019/07/04/uk-met-facial-recognition-failure-rate (accessed November 20, 2020).

11. The relative inaccuracy of face recognition for people of color, and especially women, is discussed in B. F. Klare, M. J. Burge, J. C. Klontz, R. W. V. Bruegge, and A. K. Jain, "Face Recognition Performance: Role of Demographic Information," *IEEE Transactions on Information Forensics and Security* 7, no. 6 (2012): 1789–1801, http://openbiometrics.org/publications/klare2012demographics.pdf (accessed November 20, 2020).

12. See the 2016 report to the Convention on Conventional Weapons (CCW) delegates: Bonnie Docherty, "Killer Robots and the Concept of Meaningful Human Control," April 11, 2016, https://www.hrw.org/news/2016/04/11/killer-robots-and-concept-meaningful-human-control.

13. Sex robot issues are discussed in N. Sharkey, A. van Wynsberghe, S. Robbins, and E. Hancock, "Our Sexual Future with Robots," Foundation for Responsible Robotics, 2017, https://responsible-robotics-myxf6pn3xr.netdna-ssl.com/wp-content/uploads/2017/11/FRR-Consultation-Report-Our-Sexual-Future-with-robots-.pdf (accessed November 20, 2020).

14. Health and safety issues are addressed in vulgar songs known to some groups of young men.

15. This argument is put forward strongly in K. Richardson, "Sex Robot Matters: Slavery, the Prostituted, and the Rights of Machines," *IEEE Technology and Society Magazine* 35, no. 2 (2016): 46–53, https://dora.dmu.ac.uk/handle/2086/12126 (accessed November 20, 2020).

16. Sharkey et al., "Our Sexual Future with Robots."

17. As an example of the confusion, the UK government was recently reported to be "accelerating the development of robots in the benefits system." The government is in fact referring to the deployment online of software using machine learning—nothing at all to do with robots. Robert Booth, "Benefits System Automation Could Plunge Claimants Deeper into Poverty," *The Guardian*, October 14, 2019, https://www.theguardian.com/technology/2019/oct/14/fears-rise-in-benefits-system-automation-could-plunge-claimants-deeper-into-poverty (accessed November 20, 2020).

18. See, e.g., Darrell M. West, "Will Robots and AI Take Your Job? The Economic and Political Consequences of Automation," *Brookings*

TechTank, April 18, 2018, https://www.brookings.edu/blog/techtank/2018/04/18/will-robots-and-ai-take-your-job-the-economic-and-political-consequences-of-automation (accessed November 20, 2020).

19. Amara's law, with examples, is explained in a blog by the author Matt Ridley; see http://www.rationaloptimist.com/blog/amaras-law (accessed November 20, 2020).

20. An International Federation of Robotics blog gives figures for industrial robot sales in 2018; see https://ifr.org/post/strong-performance-in-the-us-europe-and-japan-drives-global-industrial-robo (accessed November 20, 2020).

21. For a discussion of the current state of the market for cobots, see Ash Sharma, "Cobot Market Outlook Still Strong, Says Interact Analysis," *Robotics Business Review*, January 24, 2019, https://www.roboticsbusinessreview.com/manufacturing/cobot-market-outlook-strong (accessed November 20, 2020).

22. An International Federation of Robotics press release of September 18, 2019, gives figures and decomposition of service robotics sales in 2018, https://ifr.org/ifr-press-releases/news/service-robots-global-sales-value-reaches-12.9-billion-usd (accessed November 20, 2020).

23. True of the table in a MarketWatch report from September 2017, which also confuses robots with internet-based information systems and with nonrobot automation such as supermarket self-checkout machines. Sue Chang, "This Chart Spells Out in Black and White Just How Many Jobs Will Be Lost to Robots," MarketWatch, September 2, 2017, https://www.marketwatch.com/story/this-chart-spells-out-in-black-and-white-just-how-many-jobs-will-be-lost-to-robots-2017-05-31 (accessed November 20, 2020).

24. This study was testing the conversational abilities being developed for a social robot targeted at the challenging environment of a Finnish shopping mall. The one-week test, as expected, revealed some of the problems discussed in chapter 12. The false story circulated in publications from *Africa News* to the *Washington Post*.

25. This is not a purely technological issue: the Baxter and Jibo robots, both abandoned, were well-engineered products.

26. The processes through which a system maintains and reproduces itself are called *autopoiesis*, a term coined in 1972 by the Chilean

biologists Humberto Maturana and Francisco Varela to define the self-maintaining chemistry of living cells. They called autopoiesis "a network of processes of production (transformation and destruction) of components which through their interactions and transformations continuously regenerate and realize the network of processes (relations) that produced them." When autopoiesis breaks down, the system decomposes, and we die. Humberto Maturana and Francisco Varela, *Autopoiesis and Cognition: The Realization of the Living,* 2nd ed. (Dordrecht, Holland: D. Reidel Publishing Company, 1980), 78.

INDEX

Note: Page numbers in italics indicate figures.

AAAI (Association for the Advancement of Artificial Intelligence), 92, 155
Action units, facial, 170–171, 263n12
Affect. *See* Emotions
Affective empathy (emotional contagion), 174
Affordance, 26–28, 33–34, 38, 50, 94, 242n3
AI (artificial intelligence). *See also* Learning
  architecture of robots, layered, 114–115, 119–120, 157–158, 166, 253n9
  artificial general intelligence, 125
  and behavioral reactions, 115–117, 253n9
  CoBots, 120–121
  data use from sensors, 114–115
  definitions of intelligence, xiv
  emergence as a research field, 111
  "Ethics Guidelines for Trustworthy Artificial Intelligence," 222
  expert systems, 121, 254n15
  first systems, 111
  goals and programming, 122–123, 254nn16–17
  hype about, ix–xiii, 256n14
  and internet data, 122

AI (artificial intelligence) (cont.)
  and language, 211–214, 269n11
  and logic, 111–112, 124
  as mathematics, 136
  meanings of intelligence, 109–114, 125, 252n2, 253nn6–7
  noticing errors/failures, 123–125
  planning capabilities, 117–123, 254nn10–11
  and reflexes, 115
  and rules, 115–116
  sensor-driven approach to, 118–119
  skills linked with intelligence, 112–113
  superintelligence, xi, xiv, 162–163
  and symbols, 111–112, 253n4
Aibo (doglike robot), 179–181, *180*, 185, 264nn1–3
AIKON-II, 241n18
Airport buggies, 87
Albertus Magnus, Saint, 239n2 (chap. 1)
Alexa (digital assistant), 195, 204–205
Alexa Challenge, 205–206, 208
Alexandria, 3–5
Algorithms
  backpropagation, 136–137, 256n13
  bias in, xi–xii
  genetic, 128–129, 255n4
  military use of, xii
AlphaGo, xi, 92, 249n2
Alyx, 39, *39*
Amara, Roy, 227
Amazon, 204–205
Androids, 1, 18–19, 239n2 (chap. 1)
Animation, appearance of, 38–40, 244n14
ANNs (artificial neural networks), 135–138, 143–144, 159, 256nn12–14
Anthropomorphism, x
Antikythera device, 6
Ants, 146–148, 259n7
Appearance, 23–40
  and affordance, 26–28, 33–34, 38
  arms, 23–25, *24*, *32*, 32–33
  believable vs. naturalistic, *39*, 39–40, 244n14
  of cartoon characters, 38–40, 244n14
  of desktop robots, 30–31, *31*
  eyes, 27
  face, 36, 171
  and form vs. function, 23, 26
  gender, 34
  height/size, 29–32, *32*
  humanoid vs. machine-like/animal-like, 33–36, *36–37*, 38, 159, 244n11 (*see also* Uncanny valley reactions)
  of industrial robots, 23–25, *24*
  intimidating, *32*, 32–33
  legs vs. wheels, 33

lip synchronization, 38
personalization of, 28–29, *29*, 243n8
of social robots, 26–27, 242nn4–5
uncanny valley, 35–36, *36–37*, 39, 99, 177, 244n11, 250n10
Architecture of robots
drive-based, 166
layered, 114–115, 119–120, 157–158, 166, 253n9
Artificial intelligence. *See* AI
Artificial neural networks. *See* ANNs
ASD (autism spectrum disorder), 198, *199*
Asimov, Isaac, *I, Robot*, 220–221
ASR (automatic speech recognition), 207–210
Association for the Advancement of Artificial Intelligence (AAAI), 92, 155
Attentional systems, 72–73
Autism spectrum disorder (ASD), 198, *199*
Automata, 7–10, *11*, 13–14, 58, 219, 241n18
Automatic speech recognition (ASR), 207–210
Autonomous underwater vehicles (AUVs), 65, 90
Autonomous vehicles, 83, 86–90, 97–98, 133–135, 158, 230, 248nn7–8
Autonomous weapons/killer robots, 224
Autopoiesis, 272–273n26
AUVs (autonomous underwater vehicles), 65, 90
Ava (in *Ex Machina*), 18

Backpropagation algorithm, 136–137, 256n13
Baker, Kenny, 14
Bat robots, 56–57
Baxter robot, 32, *32*, 250n8, 272n25
Behavioral reactions, 115–117, 253n9
Behavioral robotics, 82, 247n2
Bert robot, 191–192
Big Dog, 244n2
Biomimetics, 55–59, 100, 147
Bird robots, 56–57
Birds, mechanical, 3–4, 240n7
Blockies, 127–128
Boids, 150
Bomb disposal robots, 26, 156–157
Boston Dynamics, 241n22, 244n2
Bostrom, Nick, 256n11
Bouchon, Basile, 241n19
Boulogne, Duchenne de, 170
Brains, human vs. robot, 135–137. *See also* ANNs
Brandeis University, 128–129
Breazeal, Cynthia, 242n5
Brooks, Rodney, 247n2, 253n9
Byzantium, 4, 58, 240n7

C-3PO (in *Star Wars*), 14, 34
CADUCEUS (expert system), 254n15

Čapek, Karel, *Rossum's Universal Robots*, 12–13, 18–19
Carnegie Mellon University, 120, 260n16
Cars, autonomous. *See* Autonomous vehicles
Cartesian dualism. *See* Mind-body dualism
Cartoon characters, appearance of, 38–40, 244n14
CCW (Convention on Certain Conventional Weapons), 225
Center for Robot-Assisted Search and Rescue (CRASAR), 157
Central pattern generators (CPGs), 48–49
Chatbots, 201–207, 212, 267n2
Chess-playing robots/programs, 13–14, 91–94, 111, 249nn1–2
CIMON robot, 161–162, 174, 176–178
Clarke, Arthur C., 3
Clocks, 3–4, 6–8, 219
Clockwork mechanisms, 5–6
Clockwork Prayer, 7
Clore, G. L., *The Cognitive Structure of Emotions*, 263n9
CMDragons, 260n16
CoBots (cooperative robots), 120–121, 152, 192
Cobots, industrial, 160, 229
Cockroach movement, 48–50
Cocktail party effect, 73
"Cogito ergo sum" ("I think, therefore I am"), 8. *See also* Mind-body dualism
Cognitive appraisal, 167–168, *169*, 170, 263n9
Cognitive empathy, 174–176
Cognitive Structure of Emotions, The (Ortony, Clore, and Collins), 263n9
Collaboration, 145–160
  box pushing, 148–149, 259n9
  cooperation/teamwork, 120–121, 152
  coordination, 152
  emergence, 146–147, 151, 159, 258n2
  flocking/formation flying, 149–150, *151*
  Game of Life, 145–146, 150
  by industrial cobots, 160, 229
  by insects, 146–148, 259n7
  and intentional stance, 159–160, 261n23
  and miniaturization, 148–149
  for search and rescue, 155–160
  soccer-playing robots, 152–155, *156*, 158, 260n13, 260nn16–17
  via stigmergy, 147–148, 152, 259n5, 259n7
  supervised, 159
  swarm robotics, 148–152, 259nn8–9
  and transparency, 159
Collins, A., *The Cognitive Structure of Emotions*, 263n9
Compliance, 45, 50–51, 96, 98
Computers as social actors, 242n4
Consciousness, definitions of, xiv

Convention on Certain Conventional Weapons (CCW), 225
Conway, John, 145
Cooperation/teamwork, 120–121, 152
Coordination, 152
Corridor, 241n22
CRASAR (Center for Robot-Assisted Search and Rescue), 157

Daedalus, 2, 5
Damasio, Antonio, 240–241n17, 262n4
Damian, John, 41
DARPA Robotics Challenge, 15, 51–52, 83, 241n20
Darwinism, 12
Data (in *Star Trek*), 18
Decision algorithms. *See* Algorithms
Deductive logic, 112
Deep Blue (chess computer), 91–92
Deep Fritz (chess program), 91, 249n1
Defense Advanced Research Projects Agency. *See* DARPA Robotics Challenge
Degrees of freedom, 42–43, 52, 100–102
Dementia sufferers, *183*, 183–184, 194–195
Deneubourg, Jean-Louis, 259n7
Dennett, Daniel, 240–241n17, 261n23
Descartes, René, 8–9. *See also* Mind-body dualism
Desire paths, 147, 259n5
Developmental robotics, 141–142, *142*, 214
Didacus of Alcalá, Saint, 7
Docklands Light Railway (DLR; London), 248n7
Doglike robots, 179–181, *180*, 264nn1–3
Dorigo, M., 259n7
Dostoyevsky, Fyodor, *A Writer's Diary*, 214
Draughtsman, 9, 241n18
Draughtsman-Writer, *11*
Dreyfus, Hubert, 256n14
Driverless vehicles. *See* Autonomous vehicles
Drones, airborne, 87, 150, *151*, 225, 231
Dualism. *See* Mind-body dualism
Durrant-Whyte, Hugh, 248n5
Dynamics, 47

EAPs (electroactive polymers), 96–97
Einstein, Albert, 255n18
Ekman, Richard, 263n13
Electroactive polymers (EAPs), 96–97
Eliza (chatbot), 201, 267n2
Emergence, 146–147, 151, 159, 258n2
Emotions, 161–178. *See also* Fear
  anger, 163, 171
  animal models of, 173

Emotions (cont.)
in animations, 172–173, 263n14
arousal and valence of, 165–166, 173, 176, 262n6
CIMON robot, 161–162, 174, 176–178
cognitive appraisal, 167–168, *169*, 170, 263n9
cognitive empathy, 174–176
via colored LEDs, 173, 263n15
and communication, 164, 170
coping behavior, 168
definitions of, xiv, 165, 262n3
disgust, 164, 171
and drive-based architectures, 166
emotional contagion (affective empathy), 174
emotional intelligence, 164
EMYS, *172*
expressive behavior, 164–165, 170–172, 174–176
facial expressions, 170–172, *172*, 175–176, 263n12
gloating, 167–168
happiness, 166–167, 171
iCat, 168, *169*, 170–171, 173–174
and intelligent behavior, 18
and machine learning, 177–178
and motivation, 163
positive vs. negative, 167
primitive, 171–172, 263n13
and reason, 162–163, 262n4
recognizing, 173–178
resentment, 167
respect/admiration, 164
robot vs. human, 165
sentiment analysis, 176–178
sorrow, 168, 170
EMYS robot, *172*
Entrainment, 49
Epigenetic robotics, 141
Equilibrioception (sense of balance), 43–44
Ethics and social impact, 217–235
of automated decision systems, 220–221, 270n3
autonomous weapons/killer robots, 224
confusing machine learning/automation with robots, 227–228, 271n17, 272n23
facial recognition's reliability, 224–225
guidelines for robotic design/manufacture, 221–224
humanlike faces, 218
humans vs. robots, 233–234, 272–273n26
jobs lost to robots and automation, 227–232, 272n25
legal codes, 221, 225
misleading people about robot capabilities, 217–219, 222–223, 231–232, 269n2, 272n24
morals of people vs. machines, 219–220
and political decisions about use of robots, 232–233

researchers' ethical principles, 223
rights of robots, 217, 219–220
robots "giving evidence," 223
sexbots, concerns about, 225–227
technologies' impact, 227–228
"Thou shalt not kill," 221
Three Laws of Robotics, 220–221
"Ethics Guidelines for Trustworthy Artificial Intelligence," 222–224
Eugenics programs, 12
Europa, 2
European Space Agency, 161
EVE (in *WALL-E*), 18
Evolutionary robotics, 255n4
Ex Machina, 18
Exoskeletons, 105–106, *107*, 108, 198, 252n21
Expert systems (encoding knowledge), 121, 254n15
Expressive behavior, 164–165, 170–172, 174–176
Eyes, pictures of, 27, 243n6

Facebook Messenger, 203
Facial Action Coding System (FACS), 263n12
Facial expressions, 170–172, *172*, 175–176, 263n12
Facial recognition, 67–70, *68*, 175, 224–225
Fear, 1–19
the Frankenstein complex, 16–17
of immigrants, 19
and robots in film vs. reality, 13–18
of robots supplanting humans, 12–13 (*see also* Jobs lost to robots and automation)
of sex robots, 1–2
of strength/invulnerability of metal figures, 2–3
and technology, 2–3
Federation of International Robosports Association (FIRA), 152
Fish robots, 57–58, 246n14
Flocking/formation flying, 149–150, *151*
Flores, Fernando, *Understanding Computers and Cognition*, 253n7
Flying robots, 55–57
Forbidden Planet, 14
Ford factory accident, 21–22, 25, 115
Fortran, 252n3
Fovea, 63, 72
Frankenstein's monster, 9, 16–17
Fujita, Masahiro, 179–180, 264n1
Functional layering, 88
Fur Hat robot, 31, *31*

Gaak, 61–62, 71
Galatea, 1–2, 226, 239n1 (chap. 1)
Game of Life, 145–146, 150
Gasbots, 76
General Problem Solver (AI system), 111, 252n3

Genetic algorithms, 128–129, 255n4
Geneva Conventions, 225
Georgia Institute of Technology, 190
Gibson, James, 26, 242n3
Global Initiative on Ethics of Autonomous and Intelligent Systems, 222
Golem, 9
Google, 196
Google DeepMind, 92
Google Maps, 77
Google Photos, 138–139, 201, 257n15
Gould, Stephen Jay, 252n2
GPS (global positioning satellites), 85, 150

HAL (in *2001: A Space Odyssey*), 17, 161
Hand of Hope, 252n21
Hanson, David, 217–218
Harvard University, 259n8
Hearing/microphones, 71–74
Hephaestus, 2
Hero of Alexandria, 4–5
Hidden Markov models (HMMs), 208
HitchBOT robot, 189
HMMs (hidden Markov models), 208
Hodgkin, Alan Lloyd, 143
Holonomic systems, 52
Homeostasis, 166
Honda, 54
HRI (human-robot interaction), 185. *See also* Social interaction
Hubris, 17
Humanoid robots
  vs. machine-like/animal-like, appearance of, 33–36, *36–37*, 38, 159, 244n11 (*see also* Uncanny valley reactions)
  rights of, 219
  soccer-playing, 153, 155, 260n17
Human-robot interaction (HRI), 185. *See also* Social interaction
Huxley, Andrew Fielding, 143
Hydraulics, 51
Hyper-redundancy, 101

I, Robot (Asimov), 220–221
IBM, 91
ICat robot, 168, *169*, 170–171, 173–174, 185
ICub robot, 141–142, *142*
IEC, 88–89
IEEE (Institute of Electrical and Electronics Engineers), 222
Illusion of Life: Disney Animation, The (Thomas and Johnston), 244n14
Inductive logic, 112
Industrial Revolution, 10
Industrial robots
  accidents involving, 21–22, 25, 115
  agency of, 25

appearance and function of, 23–25, *24*
first patent for, 228
job losses to, 229
sales of, 228–229
In-group vs. out-group thinking, 19
Insanity, defined, 123, 255n18
Insect robots, 57
Institute of Electrical and Electronics Engineers (IEEE), 222
Intel, 150
Intelligence. *See* AI
Intentional stance, 159–160, 170, 261n23
Internet of Things, 74
INTERNIST (expert system), 254n15
Inverse kinematics, 46
IQ tests, 110–111
Irpan, Alex, 256n7
Ishiguro, Hiroshi, *37*

Jacquard looms, 10, 241n19
James IV, king of Scotland, 41
Jaquet-Droz, Pierre, 9–10
Jessiko robot fish, 246n14
Jibo robot, 272n25
Jobs lost to robots and automation, 227–232, 272n25
Johnny 5 (in *Short Circuit*), 14
Johnson-Laird, Philip, 253n5
Johnston, Ollie, *The Illusion of Life: Disney Animation*, 244n14

Kalman filter, 81, 247n1
Kasparov, Garry, 91
Kaspar robot, *199*
Killer robots/autonomous weapons, 224
Kilobot, 259n8
Kinematics, 46, 116
Kismet, 38
Kohlstedt, Kurt, 259n5
Kramnik, Vladimir, 249n1
Ktesibios, 3

L3-37 (in *Solo*), 18
Language. *See* Speech/language
Lasers, 64–65, 71, 76, 84–85
Learning, 127–144. *See also* AI
and artificial neural networks, 135–138, 143–144, 159, 256nn12–14
by babies, 94, 141–142, *142*
via backpropagation, 136–137, 256n13
by Blockies, 127–128
classification systems, 139
control systems, 138
deep, 137–139, 257n15
developmental robotics, 94, 141–142, *142*
evolution as, 128–129
and genetic algorithms, 128–129, 255n4
Jenga-playing robot, 140–141
Markov decision process, 131
meanings of, 129–130
by mechanical tortoises, 143
of motor actions, 130, 139

Learning (cont.)
"paper clips" thought experiment, 134–135, 256n11
via reinforcement, 130–131, 133–135, 139–141, 256n7
Rubik's Cube manipulation, 139, 144
via simulator, 131–133, 139–140
via spiking neurons, 143–144
via statistical classifiers, 137
supervised, 130–131, 137
threat from robots that learn, 130
unsupervised, 130–131, 137
Lego Mindstorms, 254n17
Legs, artificial, 104
Leonard, John, 248n5
Lindsey robot, 186, 188
Localization, 62
Loebner Prize, 212
Logic, 111–112, 124, 253n5
Lost robots. *See* Navigation and location awareness

Machine learning, 177–178, 203. *See also* Learning
Maillardet, Henri, *11*
Markov decision process (MDP), 131
Massachusetts Institute of Technology, 58
Maturana, Humberto, 272–273n26
Metamorphoses (Ovid), 1, 239n1 (chap. 1)
Microphones, 71–74, 209
Microsoft, 201, 204, 267n1
Microsoft Kinect, *65*
Mind-body dualism, 8–9, 18, 113, 162, 240–241n17, 258n2, 262n4
Mining machines, driverless, 87, 248n8
Minsky, Marvin, xiv, 110, 253n6
Morals. *See* Ethics and social impact
Mori, Masahiro, 35, *36*. *See also* Uncanny valley reactions
Motion capture technology, 14–15, 38, 45–46
Movement, 41–59
acceleration, 47–48
and batteries/electricity, 52–55
biomimetic/bio-inspired, 55–59, 100
central pattern generators (CPGs), 48–49
of cockroaches, 48–50
compliance, 45, 50–51
degrees of freedom, 42–43, 52, 100–102
equilibrioception (sense of balance), 43–44
flying, 55–57
hydraulics, 51
jerky, 35
via motors vs. muscles, 45, 53
multiple legs, 49–50, 244n2
open-loop vs. closed-loop control of, 47–48
pogoing, 50
proprioception (kinesthetic sense), 43–45

rotation, 44, 52
sensors, 43–45
swimming, 57–58, 246n14
transfer function, 45–46
translation (change of position), 44, 52
undulation, 58–59, 246n15
and the vestibular system, 44–45
walking, 44–52
via wheels, 52, 54, 58
of Wobblebots, 50
Muscles, artificial vs. human, 96–97
Musician, 9–10
MYCIN (expert system), 254n15

Nanotechnology, 30, 148–149
Nao robot, 263n15
Nass, Clifford, 242n4
Navigation and location awareness, 77–90. *See also* Senses/awareness
by autonomous vehicles, 83, 86–90, 248nn7–8
behavioral robotics, 82, 247n2
dead reckoning, 79
via GPS (global positioning satellites), 85
via landmarks, 82–85, 87–88
location data vs. information, 77–78
via maps, 79–80
obstacle recognition and avoidance, 86–87, 89–90
odometry, 79, 84
by outdoor robots, 85
proximity sensing, 86
recalibration, 80
by robots vs. humans, 78–79
sensors for, 62
SLAM (simultaneous localization and mapping), 83–85, 156–157, 248n5
state (robot's location and velocity), 81
teleoperation, 87, 157–158, 160, 248n8
uncertainty in, 79–81
by wheeled robots, 79–81, 84–85
Neurons, 135–137, 143–144
Newell, Allen, 252n3
Northeastern University, 204
Nysa, 4

Ontogeny, 141
OpenCV (Open Computer Vision), 247n3
Origami robots, 30, 243n10
Ortony, A., *The Cognitive Structure of Emotions*, 263n9
Ovid, *Metamorphoses*, 1, 239n1 (chap. 1)

Paro (seallike robot), 182–184, *183*, 231, 264n4
Passive walking, 50
Pepper robot, *187*, 263n15
Perceptrons, 256n12
Perlin, Ken, 263n14
Philip II, king of Spain, 7
Phototaxis, 61–62
Pixels, 39–40, 63–65, 71

Planes, 56
Planetary rovers, 25–26, 124–125, 157
Planning capabilities, 117–123, 254nn10–11
Plans and Situated Actions (Suchman), 253n7
Pleo, 28–29, *29*, 243n8
Pneumatic muscles, 96
Polly World, 263n14
Programming of machines, 10, 241n19
Proprioception (kinesthetic sense), 43–45
Prosthetics, 103–106
Prostitution, 226
Proxemics, 192–194
Ptolemy II, 4
Pygmalion, 1–2, 9, 17, 226, 239n1 (chap. 1)

Quadcopters, 55–57
Quince robot, 156–157

R2-D2 (in *Star Wars*), 14, 173
Redundancy, 101
Reflexes, 115
Reinforcement learning. *See* RL
Retina, 63, 72
Reward hacking, 134
Rewards and punishment. *See* RL
Rio Tinto, 248n8
RL (reinforcement learning), 130–131, 133–135, 139–141, 256n7
Robbie the Robot (in *Forbidden Planet*), 14
RoboCup, 152, 155, *156*, 158, 181, 260n13
Robot brains. *See* ANNs
Robots
  in the ancient world, 2–5, 240n7
  becoming "superior" to humans, 103
  biological vs. mechanical, 12–13
  defined, 22, 62
  as emotionless, 18 (*see also* Emotions)
  female, 1–2, 18, 240n3
  first, 9–10
  first person killed by, 21–22, 25, 115
  hype surrounding, ix–xiii
  male, 17–18
  stereotyping of, 18–19
  videos and tests of abilities of, 15–16, 241nn20–22
  Western European attitudes to, 239n1 (intro.)
RoboTuna, 58
Robovie 2 robot, 188–189
Roomba, 123, 159
Rossum's Universal Robots (Čapek), 12–13, 18–19
R.U.R. (Čapek), 12–13, 18–19
Russell, James, 262n6

SARs (socially assistive robots), 197–198, 267n17
Saudi Arabia, 217
Scientific racism, 12
Search and rescue, 155–160

Searle, John, 212–214
Self-driving vehicles. *See* Autonomous vehicles
Self-taught robots. *See* Learning
Senses/awareness, 61–76. *See also* Lasers; Navigation and location awareness; Touch/handling
  for avoiding obstacles, 62, 68, 70–71
  cameras vs. human vision, 62–65, 246n2
  data from sensors, *65*, 66, 68–69, 247n3
  of edges, 64–67
  face/object recognition software, 67–70, *68*, 175
  for gas detection, 75–76
  hearing/microphones, 71–74
  human, number of, 43
  infrared cameras for night vision, 70–71
  for localization, 62
  for navigation, 62
  noise from sensors, 66
  numbers vs. information, 76
  optical-flow detectors, 71
  smell/electronic nose, 74–75
  for surveillance, 67, 69–70
  taste, 74–75
  taxis (sensors), 43–45, 61–62, 71, 76
Sentiment analysis, 176–178
Sex robots, 1–2, 225–227
Shakey robot, 254n10
Sharkey, Noel, 269n2
Shaw, J. C., 252n3
Shelley, Mary, 9, 16–17
Shibata, Takanori, 182–184
Shooting Star (drone), 150
Short Circuit, 14
Simon, Herbert A., 252n3
Sims, Karl, 127–128
Siri (digital assistant), 195
Skin, artificial, 98–99, 101, 250n10, 251n12
SLAM (simultaneous localization and mapping), 83–85, 156–157, 248n5
Small Size Robots, 153–155
Smart environments, 74, 89
Smell/electronic nose, 74–75
Smiles, 175–176
Snakebots, 58–59, 246n15
Soccer-playing robots, 152–155, *156*, 158, 260n13, 260nn16–17
Social impact. *See* Ethics and social impact
Social interaction, 179–199
  Aibo, 179–181, *180*, 185, 264nn1–3
  for autism spectrum disorder, 198, *199*
  butler-like robots, 192–195, 197, 199
  companion-like robots, 185, 189, 199
  for dementia sufferers, *183*, 183–184, 194–195
  doglike robots, 179–181, *180*, 264nn1–3
  expressive behavior, 192
  giving and receiving help, 192

Social interaction (cont.)
iCat, 185
by large mobile robots, 185–186, *187*, 193
limits on, 196–197
memory, 195–196
by museum-guide robots, 186, 188
natural language interaction, 196 (*see also* Speech/language)
novelty effect in, 182, 184, 186, 264n4
overtrust by humans, 190–192
Paro, 182–184, *183*, 231, 264n4
pet-like robots, 182–183, 185, 199
privacy, 195
proxemics, 192–194
rudeness/abusiveness toward robots, 188–190
rudeness by robots, perceived, 27, 194, 242n5
and social affordance, 27–28, 33–34
socially assistive robots, 197–198, 267n17
speed of robot's movement, 194
SoftBank Robotics, *187*, 263n15
Soft robotics, 100–101, 251n12
Solo, 18
Sony, 179–181, *180*
Sophia robot, 217–219, 222–223, 269n2
Specific resistance (SR), 53
Speech/language, 201–215
and AI, 211–214, 269n11
Alexa Challenge, 205–206, 208
automatic speech recognition (ASR), 207–210
of chatbots, 201–207, 212
Chinese room thought experiment, 212–214
development of, 214
error reduction vs. recognition, 210
and expressive behavior, 211
hand-coded answers to questions, 204
harassment, 205–206
hidden Markov models, 208
human-sounding voices, 210, 268n9
keyword spotting, 209–210
knowledge-based, 206–207, 268n7
language games, 214–215
microphones, 209
natural language engineering, 211–213, 215
and the robot's decision-making processes, 207
statistical approaches to, 206
text-to-speech systems, 210
Turing test, 211–212, 269n11
unit selection, 210
voice assistant for health information, 204
Speech recognition, 73
Spot, 244n2
Spy planes, autonomous, 90
SR (specific resistance), 53

SSL-Vision, 154
Stanley (autonomous vehicle), 83
Star Trek, 18
Star Wars, 14, 34, 173
Stereotyping of the other, 18–19
Stigmergy, 147–148, 152, 259n5, 259n7
Stratton, George, 246n2
Stroke, rehabilitation following, 108, 198, 252n21
Subsumption architecture, 253n9
Suchman, Lucy, *Plans and Situated Actions*, 253n7
Sugar battery, 54–55
Suitcase words, xiv, 110
Superintelligence, xi, xiv, 162–163
Surveillance, 67, 69–70, 150, 175, 231, 233
Swarm engineering, 151
Swarm robotics, 148–152, 259nn8–9
Swimming robots, 57–58, 246n14
Symbol grounding, 253n4

Talos, 2, 5
Taste, 74–75
Taxis (sensors), 61–62, 71, 76
Tay (chatbot), 201, 203, 267n1
Teamwork/cooperation, 120–121, 152
Technology
  and fear, 2–3
  motion capture, 14–15, 38, 45–46
  nanotechnology, 30, 148–149
  social impact of, 227–228
  special effects, 14–15
Telemetry suits, 14
Teleoperation, 87, 157–158, 160, 248n8
Ten Commandments, 221
Terminatrix (in *Terminator 3*), 18
Termites, 146–147
Tesla Autopilot, 88–89
Tetraplegia, 106
Textile looms, 10, 241n19
Theory of mind, 174
Thermostats, 23, 25, 219
Thinking, definitions of, xiv
Thomas, Frank, *The Illusion of Life: Disney Animation*, 244n14
3D printers, 128–129
Three Laws of Robotics, 220–221
Tin Man (in *The Wizard of Oz*), 14, 34
Touch/handling, 91–108
  arms with range-finding sensors, 97–98
  and artificial skin, 98–99, 101, 250n10, 251n12
  by chess-playing robots, 91–94, 249nn1–2
  compliance, 95–96
  and exoskeletons, 105–106, *107*, 108, 198, 252n21
  grippers vs. human hands, 101–103
  hand-eye coordination, 95
  hugging, 100
  human sense of touch, 98

Touch/handling (cont.)
for kitchen tasks, 94
motor tasks, 91–92
object recognition and grippers, 94–96
and obstacle avoidance, 97–98
pancake-flipping robots, 93, 249–250n5
physical contact with humans, 97–100, 250n8
pneumatic muscles, 96
and prosthetics, 103–106
robot vs. human capabilities and knowledge, 93–94
segmented arms, 93
shaking hands, 100
shape memory, 96–97, 101
soft robotics, 100–101, 251n12
and soft sensors, 99
and touch screens, 98
Tractors, driverless, 87
Trains, driverless, 87–88, 90, 248nn7–8
Transfer function, 45–46
Transhumanism, 103
Turing, Alan, 211–212, 269n11
Turk (Mechanical Turk; Automaton Chess Player), 13–14
Turriano, Juanelo, 7
Twitter, 201, 203
2001: A Space Odyssey, 17

Uncanny valley reactions, 35–36, *36–37*, 39, 99, 177, 244n11, 250n10
Understanding Computers and Cognition (Winograd and Flores), 253n7
United Nations, 225
University of California at Davis, 205
University of Cambridge, 133
University of Hertfordshire, *199*
University of Washington, 264n2
USAR (urban search and rescue), 155–156
US Army, 108

Vacuum-cleaner robots, 30, 123, 159, 185, 219, 221, 230
Varela, Francisco, 272–273n26
Vestibular system, 44–45
Vision, human
vs. cameras, 62–65, 246n2
inverting glasses, 246n2
seeing without registering, 72–73
Vygotsky, Lev, 269n14

Walking, 44–52, 105–106
WALL-E, 18, 172
Walt Disney Company, 194, 217–218
Walter, Grey, 143
Wasps, 147–148
Watson (question-answering system), 177
Weather systems, 146
Weizenbaum, Joseph, 202, 267n2
Westinghouse, ix–x

Williams, Robert, 21–22, 25, 115
Winograd, Terry, *Understanding Computers and Cognition*, 253n7
Wittgenstein, Ludwig, 214
Wizard of Oz (human-robot interaction schema), 15–16, 241n21
Wizard of Oz, The (film), 14, 16, 34
Wobblebots, 50
Wright brothers, 41
Writer, 9–10
Writer's Diary, A (Dostoyevsky), 214

XCON (expert system), 254n15

Yale University, 101, 251n12

Zeus, 2
Zombies, 35